The Dog Wars

How the Border Collie Battled the American Kennel Club

Donald McCaig

OUTRUN PRESS

My research into sheepdog and dog fancy history was abetted by:
Dean Douglas Gordon
Ms. Lucia Stanton
Ms. Evelyn Timberlake
Ms. Mary Thurston
Ms. Penny Tose
And the librarians of The Alderman Library, University of Virginia; the Chapin Dog Collection at the College of William and Mary; The Department of Agriculture Library at Beltsville, Maryland; the Virginia Tech Library; the Library of Congress; and the Kennel Club Library in London.

Cover design by Denise Wall.

ISBN: 9780983484509

Library of Congress Control Number: 2007925955

DEDICATION

What began as an ugly bit of dog politics became for me a journey of discovery. My account may leave the impression that I was more instrumental in the Dog Wars than I was. In truth, I was only one of the Border Collie community who overwhelmingly rejected the American Kennel Club's scheme to reduce a brilliant working dog to a handsome nitwit.

Too many people from too many dog breeds gave time, money, and energy to this fight to thank them all by name. Some who helped us are inside the AKC hegemony and would not thank me for mentioning them.

It was a great privilege to know and work with people so willing to stand by their dog.

This book is dedicated to them.

Argus' form is good, but I am not sure if he has speed of foot to match his beauty, or if he is merely what the table-dogs become which masters keep for show.

Homer, *The Odyssey*

Contents

1

Pip

Oh, where have you been,
Pippy-Boy, Pippy-Boy?
Oh, where have you been,
charming Pippy?
I have been to the State Fair
and I won a ribbon there.
But I'm a young dog and cannot
leave my Farm

In the spring of 1991, we were rich with dogs. Eight years before, I'd bought a sheepdog pup, sight unseen as a birthday gift for my wife, Anne. Soon enough to justify Anne's skepticism about husbandly gifts, I started training him.

Border Collies can puzzle out routine farmwork without much help, but training and working a sheepdog to anything like his full capacity is a profession that takes years to learn. I was, and am, a big man. I was also a loud man. Fortunately, Pip was as hard as I was loud and when I bellowed at him, Pip'd give me a hot look meaning "I am more dog than you deserve" before he'd regroup and give our work another try.

If Pip had been a soft dog I would have ruined him and my life would have taken a different course.

As Anne can (and will) attest, I am a flibberty-gibbet, a man of short, hot enthusiasms. After I finished a novel about Border Collies, *Nop's Trials*, I thought I was done with dogs, and it was almost two years before I understood how deeply sheepdogs had nestled into my life. I bought Silk as a wife for Pip, Mack was their son, Gael was the bitch I searched Scotland to find (the wee heroine of my *Eminent Dogs, Dangerous Men*), and Harry was Gael's son.

We operated a 280-acre sheep farm in the western mountains of Virginia. When lambs were born, we used our dogs to bring them and their mothers into the barn. Dogs fetched sheep for feeding, worming, shearing, and foot trimming. Our dogs sorted sick sheep from the flock and brought them to the pen for doctoring. When a neighbor's bull smashed through our wire fence en route to our milk cow, dogs dissuaded him from his amours. Night or day, I never went to the livestock without a dog at my side.

We didn't own wall-to-wall carpets; our floors were bare wood. Once, in an optimistic moment, Anne bought a charming little Swedish throw rug, but it wasn't long before it went to the dumpster. When a dog's been indoors too long, charming Swedish throw rugs look mightily like grass.

Every bit of furniture within puppy reach looked like beavers had gnawed, and we repainted scratches on our kitchen door every spring.

Once a week we swept up enough dog hair to create a small dog *ex nihilo*. UPS men approached our house cautiously.

Anne called our dogs, "The Gang of Five."

In the spring of 1991, lambs fetched sixty-nine dollars a hundredweight at the market, and our fall-seeded alfalfa had overwintered successfully. Uniform dark green carpeted our hill field. Our living was sheep farming (Anne was shepherd) and writing (me) and that spring we weren't broke or in debt—an unusual turn of affairs. By my reckoning, I wouldn't have to go to the bank until early July, and who knows what miracles might happen before that? By one such miracle, *Eminent Dogs, Dangerous Men* was officially a best seller.

The ABA (American Booksellers Association) had held its meeting in New York City, and in its fat convention issue, *Publishers Weekly* listed the host city's best sellers. As it happened, PW took its list from a *Village Voice* survey of New York's hip literary bookstores. Since the survey source was noted in disclaimer type and since there *Eminent Dogs* was, big as life, number seven on PW's list, I fantasized that when all those booksellers returned home they'd promptly order tens of thousands of copies. I mean, really, who can ignore a New York City best seller?

Eminent Dogs got kindly reviews, which I forwarded to my mother. Naturally I thought this modest hubbub was long overdue, but only a Grub Street newcomer would expect a book about plain Scottish shepherds and their brilliant sheepdogs to sell like the book that enthralled the media at the time, Julia Phillips' *You'll Never Eat Lunch In This Town Again*, which discussed glamorous Hollywood agents, their life philosophies, and drug ingestion.

♠♠♠

In 1991, I knew all the sheepdog handlers. Though I might not know their livelihoods or how many kids they had — or even if they *had* kids — I knew their dogs, and the sheepdog dreams they didn't tell me I could guess because they were the same as mine: one day, at one trial, my dog would show everybody on earth just what a wonder he was.

Sheepdog people were excluded from the greater world of American Kennel Club purebred dogs (the "Dog Fancy"), but no sheepdogger cared.

From time to time, an AKC "herding dog" was signed up for a sheepdog clinic. These were perfectly nice dogs, well groomed, well cared for, and their owners loved them. The poor beasts would run around the training ring looking everywhere but at the sheep, or they'd take a pee, or they'd bark or, sensing that something was expected of them, they'd start doing pet tricks. Poor dog, poor owner.

For centuries, Border Collies had been bred exclusively for abilities. Though some — like the Gang of Five — lived indoors, few were well groomed, and some spent their non-working hours at the end of a short chain. Big ones, little ones, handsome ones, ugly ones, long-coated, short-coated: nobody gave a damn. How's his outrun? Can he read sheep? Can he move a rank old cow? In the rough ordeal of the twentieth century, only a few breeds other than the Border Collie were still bred for performance: Arctic sled dogs (the ones that run the Iditarod); fighting pit bulls (though there can't be many of these still being bred); some bird dogs (like the English setters that trial at Ames Plantation, bred for hunting abilities, who are not the same breed as those "English Setters" registered by the American Kennel Club). Although the dogs Westerners use to chase down and kill coyotes have been bred for performance, I'm not sure they are a "breed."

Over countless dog generations, the shepherd's everyday work determined the skills bred into the sheepdog. Sheep are most profitable on marginal land, and in Britain most sheep grazing is rough moors, fens, or steep lichen-covered hills. A typical Scottish shepherd will have charge of a thousand ewes scattered over as many acres. That shepherd and his dogs must gather the sheep, treat the injured and sick, push the sheep through the tanks where they're dipped for sheep lice and keds. Man and dog must catch frightened ewes with part-born lambs dangling behind, extract randy rams from ewe flocks, and separate ewes from their weanlings.

The shepherd's dog must be intelligent, trainable (not the same thing), and possess those specialized instincts only sheepdogs need. The dog must be obsessive. Its work must mean more to the dog than sex, food, or ease. Many a bitch will abandon her puppies to work.

Every now and again, in hot weather, someone carelessly works his Border Collie to death.

I have never met a sheepman — and some are rough, hard-handed characters — who wasn't grateful to his dogs. Over the years, the Gang of Five had saved our flock from flood, and blizzard, and had retrieved escaped sheep from the county road our most dangerous night of the year: Saturday night before opening day of deer season.

Evening chores in winter; the failing light gave a blue cast to the snow and the farmhouse seemed a world away. My life was me and my dog and the work we shared.

♠♠♠

Nobody visited our farm without a sheepdog demonstration. The Gang loved to show off; when strangers appeared their tails fluffed and their eyes brightened: "I'll do it best! Me, pick me!" Pip was particularly fond of girls, so when a zoology teacher brought a dozen college girls to a real working sheep farm, I couldn't deny the old boy his chance.

In the barn the girls ohhed and ahhed over our newborn lambs. When I handed one a lamb, she didn't quite know what to do with it: how do I hold him, how tight can I squeeze? One girl asked, as someone always does, whether we raised our lambs for wool or meat and was distressed to hear these lambs would eventually go for slaughter. One girl shuddered and wore a virtuous vegetarian expression.

It had rained during the night and the sky was cornflower blue with only a couple smug clouds. The grass was spring green.

Pip was the Maurice Chevalier of Border Collies; he poured on the charm, swirling around the girls' feet, grinning handsomely.

If they think about dogs at all, most people believe dogs are dumb, loving, loyal chowhounds. Sure, you can trust them around your kids, but you can't trust them to come when called. Border Collies who take whispered split second commands several hundred yards away seem like a strange new species to those who expect so much less of their dogs. Pip knew applause was guaranteed, and he warmed up his audience like a trouper.

The sheepdog runs out, gathers the sheep, and brings them back to the shepherd. This is a genetic modification of a wolf strategy: the small, fast wolf which runs out to turn prey animals back into the jaws of its slower, more powerful, packmates. The Border Collie doesn't bark or snarl, or bite at the sheep. Instead it crouches and "eyes" them; symbolically, he's a predator with a plan.

It is possible to train a Border Collie to shun sheep, and Search and Rescue Border Collies in Britain are so trained. By contrast, if you're no dog trainer, just an overworked sheep or dairy farmer, you can take your young Border Collie to your stock, say, "Shep, go get 'em," and Shep will (in a rough-and-ready fashion) "go get 'em." Often, people come up to me at trials to share memories of beloved dogs: "Shep'd bring in the cows, morning and evening. It was like he could tell time. We never trained him. I guess he just trained himself."

No other extant dog breed can run out, gather stock, and fetch them from a great distance, and, of course, no wolf can either.

The day of Pip's demo, our sheep were waiting beneath an ancient oak in a forty-acre pasture, three hundred yards from his schoolgirl audience.

Pip set out nicely from my feet, running smoothly until — two hundred yards out — he stopped dead, jumped straight in the air, spun in a tight circle, jumped up again, paused, and resumed his outrun.

Pip hadn't told me he was going to do that. Before Pip left my feet he always told me where he was headed and how he felt about his work today. I, in turn, would tell Pip where the sheep were (if he couldn't see them) and something about the nature of his work. If this was an emergency situation, Pip expected me to let him know. If he got lost on the way out, he expected me to correct him.

Our conversation was body language, eye contact, and murmurs. If the sheep are among cows (a difficult situation for the dog) I cannot inform Pip of that specific hazard, though I can urge him to be alert. Although it is easy to excite a dog, if there is any way to calm him before he's sent, I have yet to discover it. The sheepdog is perfectly indifferent to any last minute avowals of affection which are more likely than not to throw him off stride.

Running in a circle—jumping straight in the air: I'd never had any dog do that before. I feared stroke, seizure; maybe Pip'd stepped on a copperhead.

My snakebite guess was supported by Pip's subsequent moves: he came behind his sheep very slowly, like an old, very sick dog. Pip's head was low to the ground, almost dragging.

The girls couldn't know that normally the sheep would be coming at a gallop and I'd be whistling to slow Pip down. They thought any dog working sheep was pretty neat. They applauded.

When he came near enough, I saw Pip had a woodchuck in his mouth, a big one, maybe eight pounds, a fifth of Pip's body weight. His little dance out there had been the danse macabre. The unlucky chuck had popped out of his hole just as Pip came flying by, and on the instant, Pip decided to ad lib a new finale to his performance: "Mister Pip! Ta, da! Not just a sheepdog, not just a charmer! Pip, the varmint dog!"

Pip quit his sheep to drop his prize, triumphantly, at the vegetarian's feet. Thud. She did not respond as Pip had anticipated.

2

A Brief History

Thomas Jefferson did the first recorded sheepdog demonstration, and as late as 1848 agriculturalists were still snickering. As they told the story, Mr. Jefferson imported his sheepdog (a beast somewhat like the modern Briard) from France, and one fine evening invited his dinner guests onto the broad terrace behind Monticello where his Merino sheep (imported from Spain) were grazing peacefully. Perhaps Jefferson and his guests had drunk beaucoup wine.

Jefferson said, "Go get 'em, Shep!" ("*Bon Chance, Bergere!*"), whereupon the dog ran at the sheep like a bat out of hell, and the sheep bolted for the nearest precipice where they fell to their deaths.

Good story: untrue.

In 1789, when Jefferson returned from France, he brought a "chienne bergere" in whelp, not because he had work for the dog — the first Merinos didn't arrive in the United States until 1809 — but because Buffon, the naturalist Jefferson most admired, asserted that the sheepdog was the highest type of dog, the "true dog of nature, the one . . . that must be regarded as the root and model of the entire species." Jefferson promptly added the shepherd's dog to the list of old world animals that should be introduced to the new world. Contrary to later stories, Bergere (and most of her pups) thrived, and she was especially clever at putting chickens into their roost at dusk.

In 1809, Lafayette shipped Jefferson a fully-trained sheepdog, and Thomas Jefferson became a sheepdog convert, providing offspring to other agricultural improvers in Pennsylvania and Kentucky. Jefferson believed his dogs were "the most careful, intelligent dogs in the world."

These sheepdogs were drivers, not fetchers, black in color, some thirty inches high, and one of Jefferson's slaves recalled years later that the dogs had stumpy tails.

Jefferson's French sheepdogs weren't America's first. Spanish sheepdogs arrived with Coronado's pastors in 1540, and one writer places their descendants in Mexico in the 1570s. In 1841, when George Kendall saw them in Sante Fe, Coronado's dogs (probably mixed with Native American dogs) were called "Mexican Sheepdogs":

> There was no running about, no barking or biting in their system of tactics; on the contrary they were continually walking up and down like faithful sentinels on the outer side of the flock and should any sheep chance to stray from its fellows, the dog on duty at that particular post would walk gently up, take him carefully by the ear and lead him back to the flock. Not the least fear did the sheep manifest at the approach of these dogs.

Any sheepman who has ever taken hold of a sheep's ear has cause to question this account (sheep resent ear gripping), but Kendall was a keen observer and we should give him the benefit of the doubt. Here's an 1836 account:

> A flock was feeding near the road, attended only by a large dog, and one of the men . . . fired his musket at one of them and shouldered it (the carcass). At the crack of the gun the dog ran around the scattered flock, herded them in an instant, and then made chase after the man and forced him to drop the sheep.

Also this, from a bit later:

> During the seventies the master of an unusually clever sheepdog was killed by the Indians in Colorado. For some reason the sheep were not taken. But the dog continued to perform the shepherd's duties, keeping them together and guarding them from predators and at night rounded them up and drove them into the great corral belonging to that range, and lay in the gateway all night, since he could not close and fasten the gate. Two months after the raid, the owner (of the flock) found them to his great surprise, under the care of the dog, not one missing.

These accounts describe one dog doing what today we employ two very different sorts of dogs to do: the sheepdog (Border Collie, Australian shepherd, Kelpie, McNab) and the sheep guardian dog (Anatolian shepherd, Maremma, Navajo dog, Akbash).

The sheep guardian pup is bonded to sheep, reared with and *as* a sheep (some Mexican sheepdog pups were even taught to suckle ewes or goats). The modern sheep guardian dog is unresponsive to human commands, and though it accompanies the flock and drives off (or kills) predators, it cannot fetch sheep, drive them, or bring them in at night.

Excepting their contributions to the Navajo dog gene pool (and remotely, the Catahoula's), these Spanish (Mexican) sheepdogs are extinct in America, replaced by the modern guardian breeds.

As there were many British sheep varieties (Suffolk, Dorset Down, Wiltshire, Scottish blackface, Welsh mountain), so there were many British collies (the Welsh gray, the Dalesman, the Wicklow collie, the Smithfield collie). Some writers have attempted to give collies a venerable history — supposedly they arrived in Britain with the Romans or perhaps the Norsemen — but since modern livestock grazing practices didn't commence in Britain until wolves were eradicated in the seventeenth century, these attempts to provide an ancient lineage are unconvincing. The name "collie" is derived from "coal colored." They were poor men's working dogs whose owners had no more interest in "purebred dogs" than the modern California rancher who routinely crosses different breeds of working sheepdogs.

Nineteenth-century British collies weighed between twenty-five and seventy pounds, had short or long coats, flop or prick ears. Some were predominantly black and white, some were brown and white or yellow and white and some reddish in color. What all collies had in common was genetically fixed abilities. They would all run to the head of sheep, gather a flock, and fetch them to their shepherd. They were all biddable (trainable) and keen to work.

Queen Victoria was introduced to collies at Balmoral in 1848 and brought several to London where, soon enough, they became the rage. Dog fanciers claimed their show dog was the true and only collie and came in just two varieties, based on coat: Rough and Smooth.

Later, other British collies (a. k. a "working collies") began to be registered by an agricultural association, the International Sheep Dog Society (ISDS) , whose avowed purpose was "to secure the better management of stock by improving the shepherd's dog."

In 1924, the London Daily Mail invited the ISDS to put on a sheepdog trial in Hyde Park where the most able "collies" in Britain would compete. The (British) Kennel Club, which had been registering show collies for decades, objected: "Why, these aren't collies at all," they said.

"Very well," ISDS Secretary Reid replied, "We'll call ours *Border* Collies."

And a new name for the working breed was born.

3

The Border Collie in the United States

B y 1850, farm collies were common in the northern United States, and three varieties became breeds in America: the McNab (California, 1858), the English shepherd (East Coast and Midwest, 1890s) and the Australian shepherd (West, early twentieth century).

The first American sheepdog trial was held in Philadelphia in 1888.

Arthur Allen was the only expert, articulate dog man who knew both the nineteenth-century stockdog and the modern Border Collie. As he wrote in his *Lifetime with the Working Collie*, the nineteenth-century dogs "were large and more stern than the Border Collie of today and were used for working all kinds of livestock. A descendant of these dogs that my Father owned was Old Cast and was so stern a dog that you never could pet him, but was as reliable and faithful in his work as any dog he ever owned."

In 1900, when Allen's father and Old Cast were hired to drove mules to Tennessee, his "dogs were used to contain the mules and were also very important as guard dogs because there were many highwaymen waiting to prey on travelers."

In 1908, plant breeder W. A. Burpee sold farm collies from his Fordhook Kennels to farmers served by railway express. You could pick up your new working partner at the station along with your baby chicks and garden seeds. Though disparaged by show breeders, these farm collies were wonderful, intelligent, useful dogs.

Early American sheepmen were rough, practical agriculturalists. Most respected their dogs, and some admitted spiritual kinship with them. But first and foremost their dogs were tools, and useless dogs were driven off or shot.

In the twenties, Scottish shepherds like Sam Stoddard and Tom Bradburn brought Border Collies with them to America and were soon putting on sheepdog demonstrations at livestock exhibitions and rodeos. Arthur Allen did demonstrations, handled his dogs in two Disney movies (*The Arizona Sheepdog* and *Nick, A Sheepdog*) and for many years traveled with the Roy Rogers rodeo. Unless they'd been hired to do a demonstration at a dog show (as Carl Shaffner once did at the Westminster Kennel Club Show) few handlers had ever seen a dog show and I know none who'd ever showed at one.

Sheepdog culture in America shared the mores of rodeos and livestock exhibitions. The constitution and bylaws for the United States Border Collie Handler's Association, founded in 1979, were based on the bylaws of the Rodeo Cowboys Association, and the traditional farewell at the end of a sheepdog trial is the cowboy's "See you on down the road."

Because early handlers owned and trained their own dogs, the professional handler who kept and showed dogs for rich absentee owners never found a niche in sheepdog culture. Achievement other than achievement with the dogs was devalued, and I couldn't guess at the occupations of many handlers I've known for years. In this proletarian culture wealthy handlers often conceal their affluence. Because training and handling skills were (and are) revered, no person can aspire to any position within sheepdog organizations unless he (or she) has demonstrated uncommon skill with the dogs. Imagine an NBA where every owner was a great player, an Olympic Committee comprised entirely of gold medal athletes, and you'll understand the advantages (and disadvantages) of such an arrangement. Meetings of these organizations were held in conjunction with trials, because that is the only way to get a quorum. (George and Topsy Conboy celebrated their fiftieth anniversary with a sheepdog trial, and if I die during the prime fall trialing season, I hope my executor will have the good sense to put on a Funeral Trial.)

In sheepdog culture speech is laconic, and praise for man or dog understated. It can be funny to watch the newly sheepdog obsessed adapt to that culture that nurtures their dogs. As his (her) dogs improve, many a previously garrulous suburbanite starts to mutter like John Wayne.

♠♠♠

In the 1960s, there were two North American registries, Arthur Allen's North American Sheep Dog Society (NASDS) and Dewey Jontz's American International Border Collie registry (AIBC, an offshoot of the NASDS). When I say "Arthur Allen's" and "Dewey Jontz's," I mean it. These registries were operated by and for their owners. Though each had Directors, the Boards were self perpetuating and largely honorific, and no "member" had much say. Dog registries whose services are keeping a studbook and exchanging pieces of paper for money are lovely rural businesses.

The United States Border Collie Handler's Association was an ad hoc group of handlers who met at sheepdog trials. The Pulfer brothers, Topsy and George Conboy, David Rogers, E. B. and Francis Raley, Bud Boudreau, and John Bauserman were its most active members. Their first National Finals, hosted by a man who'd never run a dog and judged by men who hadn't trialed in years, was a far less important trial than the North American (Arthur's trial) or the Kentucky Bluegrass, held at Lexington's Walnut Hall.

There were a handful of other Border Collie Associations: Virginia, Texas, Oregon, the idiosyncratic Redwood Empire, Alabama, and the New England (now North East) Border Collie Association (NEBCA).

♠♠♠

Trials in these early days were few and far between. There were ranch trials in Texas, state fair trials in New England and Virginia. On the West Coast, the Redwood Empire Club held three or four fairground trials. There were a few trials in the deep South. Maybe, in all, forty sheepdog trials in North America. The standard of work was poor. Most trials were "timed" trials (based on speed rather than on accurate and correct work), in which the sheep were contained in a wire enclosure which was collapsed or opened when the dog arrived to work them. Most trials didn't have a shed (in which one sheep is separated off from a group), because few handlers knew how to shed.

Many Brits saw us as a high dollar market for their culls, and the American handler who imported dogs was sometimes favored by the Brit who judged that handler's trial.

Unless you were lucky enough to neighbor a good handler, there was no practical way to learn how to train or handle a dog. You'd buy a pup from the big hat who did a demonstration at your county fair, and six months later when your pup started to work the big hat was halfway across the country. There was no Border Collie magazine. In Britain, Matt Mundell published a short-lived mimeographed sheet, and Sheila Grew began publishing the *Working Sheepdog News*. Richard Karrasch, Chuck O'Reilly, Carl Shaffner, Pope Robertson, and Foy Evans had written training guides (available from the authors or the Raleys' mail order bookstore, which also carried Longton and Hart's *The Sheepdog: Its Work and Training* and John Holmes's *The Farmer's Dog*).

Women competitors were barely tolerated: Betty Levin, Inez Schroeder, Lena Bailey, Ethel Conrad, and Ada Karrasch (who wore a bright red western jumpsuit and cowboy hat onto the field because the Good Old Boys hated female flamboyance).

Border Collies were starting to be popular in AKC obedience, but few dog fanciers and most Americans had never heard of the dog. In 1978, with little opposition from the ISDS, the Kennel Club (UK) had "recognized" the Border Collie for conformation showing. The Australian National Kennel Club had recognized the breed in 1963.

♠♠♠

Arthur Allen was a huge presence. The only man in America making a living from Border Collies, he'd won the North American Championship many years in a row and did sheepdog demonstrations at the major stock shows, including the San Francisco Cow Palace. Although I never saw Arthur work, I have seen his Disney short, *The Arizona Sheepdog*. He'd be a hard handler to beat today. Scottish handlers told me, "Oh, he'd come over for the International and he'd take notes on every dog, and God help the man who went up to Arthur Allen during a run."

Allen was not known for kindnesses to his competitors. When one rival announced a trial, Arthur went to the man's sheep supplier, bought the sheep, and held his own trial.

In 1981, Arthur began tangling with another Border Collie giant — on the surface of it, an unlikely one. Bill Dillard was a bachelor retired teacher. Bill was no great handler, but he had time, brains, and a passion for the Border Collie. David Rogers says that if anyone came to Bill "talking like they *might* even be interested in hosting a trial, Bill would talk them into it." Bill was responsible for a new circuit of trials through Mississippi, Alabama, and Georgia, and he encouraged trial hosts in Tennessee. He published *The Southern Stockdog Journal*, North America's first stockdog magazine, and got on the board of Allen's NASDS, where he pestered Arthur seeking a more democratic registry.

Everybody was hoping the AIBC and NASDS would merge. Fred Bahnson came from Winston-Salem money, had been a North Carolina State Senator, loved sheepdogs, and had sponsored Jack Knox's immigration to America. President of the NASDS, Fred was as tough and wily as Arthur,

The 1982 NASDS Board of Directors meeting was held at the Bluegrass trial. Most of the Directors had sent proxies to Arthur, including Fred Bahnson, who was vacationing in Italy. News reached the trial that Fred had had a serious stroke and had to be flown home in a hospital plane. Arthur held the Board meeting anyway, and with his proxies, drummed the dissidents off the Board.

Shortly afterwards, rumors started that a new registry was being formed. We wondered: "Why do we need a *third* registry?" Nobody quite knew who was promoting this registry, and until Bill Dillard published the American Border Collie Association's bylaws in the *Southern Stockdog,* nobody knew what it stood for.

It stood for a democratic, member-owned registry: one member—one vote. Leroy Boyd, Bill Dillard, and Ralph Pulfer were officers, and working at night after her full-time job, Patty Rogers was the ABCA Secretary.

That 1982 Bluegrass had another important consequence: the young, talented handler Bruce Fogt won with his bitch Hope. More and more, younger handlers were entering trials and doing well. Friday afternoons, Bill Berhow would put his bitch Scarlet on the back of his motorcycle and drive eight hours from Florida to Bill Dillard's, where they'd work dogs until Sunday night. Kent Kuykendahl decided he was more interested in sheep dogs than the sheep his family was famous for. Cheryl Jagger, who'd grown up with her father Walt's Border Collies, started giving clinics.

These keen, competitive younger handlers started beating men who'd previously been unbeatable. Their clinics taught hundreds of new handlers.

They raised the bar.

In 1984 my *Nop's Trials* was published. While there'd been many fine children's and young adult books about Border Collies, there'd never been an adult novel. Heavily promoted, Nop introduced several hundred thousand readers to Border Collies. Its virtue was the warning on the last page that "Border Collies do not make good pets," a warning the community has repeated often enough it has reached the general public.

Trials and clinics exploded. One January, Jack Knox gave a demo at Montana's Winter Livestock Show. He was invited to return in the spring for a clinic. That fall, the Montana Stockdog Association formed and had its first fun trial.

Most trials were now judged, most new trials were on national-style courses, and many had dog food sponsorship. Purina sponsored important trials, a circuit championship, and the National Finals. The AIBC had suffered when Dewey Jontz died and when Dewey's heir, Dean Kaster, died, its members flocked to the ABCA. By 1991, the premier North American trial was the National Finals, and the best U.S.

handlers were as good as the best handlers in the UK. Some of those top handlers were women.

♠♠♠

Sheepdog trials are not self-referential: they are designed to produce dogs useful in the practical world. Nor are they rule bound. The ISDS *Rules for Sheepdog Trials* covers one side of a piece of paper, and the principal catechism (J. M. Wilson's *Notes for Judges*) covers the other side. I can hire anyone to judge my sheepdog trial and lay out the course however I wish.

The impetus is toward greater difficulty. When sheep were penning too easily at New York State's Leatherstocking Trial, judge Tommy Wilson drove a stake fifteen feet away from the pen where the handler had to stand, just to make everything harder. But if I were to include a pet trick in my trial—like fetching a frisbee or rolling over and playing dead—my peers would be appalled.

I can't remember when I last petted someone else's dog at a sheepdog trial. Barking dogs are culturally offensive, and when one does bark (usually recognizing its owner's whistles on the field), people wince. Dogs coming on or off the field are rarely leashed, nor are those dogs behind the fence watching the action. Before and after the trial, gangs of Border Collies run around and play. In a quarter century, I have never seen a dog fight at a sheepdog trial.

Sheepdog trialing does not attract many young people, but handlers in their seventies are unexceptional. Several men have suffered fatal heart attacks at the handler's post and I wouldn't mind going that way myself, though it must baffle the dog.

♠♠♠

While I was busy doing everything else life requires, Pip got old. He'd learned all there is to know about sheep and too much about me. When push came to shove, he'd ignore my commands (suggestions?) and get on with the job he understood perfectly, thank you so much.

The Seclusival Trial is two hours drive from our farm, and I'd left home late. As soon as I arrived, I jumped Pip out. Nobody was at the handler's post, so I asked Lyle Boyer who was up.

"You are," she said.

Since trial sheep behave differently throughout the day, running order is decided by lot. At smaller trials, if you arrive late, the trial director may accommodate you, but at a big trial if you miss your turn, you won't get a second chance. I jogged toward the course asking Lyle about the judge's instructions.

"Left-hand drive," Lyle said. "Pen before you shed."

"Sorry," I called to the judge as I walked onto the field. I sent Pip casually, before I had even reached the handler's post.

Although there must have been informal competitions earlier, the first modern sheepdog trial was held at Bala, Wales, in 1873. The historian and sheepdog handler Albion Urdank notes that this trial was intended as a country entertainment—aristocrats enjoying bumpkins at play. But sheepdog trials were swiftly appropriated by the bumpkins and their patrons, agricultural improvers who decided that trials should not seek friendly, pretty, aristocratic, or even competitive dogs. Instead, they sought dogs that would make it possible for a man on foot to handle a thousand sheep on mountainous, unfenced ground, dogs that could work on their own, take whistled instructions from over a mile away, and travel a hundred miles a day in the foulest weather without complaint. That's what trials are for: to choose the sires and dams of the next generation of sheepdogs. They are a paradigm of the dogs' daily work, made more difficult.

As a tool of genetic selection, the sheepdog trial has done exactly what its creators had hoped. Very few Border Collie pups won't work stock. That's not to say that all Border Collie pups will grow into first-class dogs, or even that they'll all make trial dogs. There are timid pups and stupid pups and neurotic pups and pups too fey or willful to take training. But, for all that, given the slightest chance, most Border Collies will work stock. And that's why people remember old Shep, "He just trained himself."

Since the sheepdog trial functions as a winnowing device, it would be counterproductive to restrict entries. Any dog can enter an Open trial, any age, any registration (or none), any breed—although if you bring your wolfhound to my trial I shall ask to see him work a few sheep in a small paddock before I turn him loose on my sheep in a big open field. After twenty years and several hundred sheepdog trials, I have seen three bearded collies (crosses?), one Australian cattle dog and one Shetland sheepdog run in a sheepdog trial. I am told there was an Australian shepherd (cross?) out West who was competitive, and in 1996 and 1997 Butch Larson qualified a Kelpie to run at the National Handler's Finals. All the other dogs at the trials were Border Collies, and though most were registered, some were of unknown parentage from the animal shelter.

"This dog over here—what is it?"

"Border Collie."

"And this one, with all the funny colored splotches and—look, it's got a blue eye."

"Border Collie."

"If they're all the same breed, why do they look so different?"

This question is so persistent most sheepdog handlers have a stock answer. When a reporter asks me what Border Collies are supposed to look like, I reply, "What does a reporter look like?"

They don't look alike, they're alike in what they do, and the sheepdog trial serves as the Border Collie licensing board, review committee, law board, tenure committee, FAA, FCC, and admissions committee for sheepdogs.

As Gerard Manley Hopkins put it: "For what I do is me! For that I came!"

That day in May, when Pip and I walked onto the course at Seclusival, we had a hundred points. As Pip ran, the judge would deduct points for every mistake. The judge cannot award points for style or cheerfulness or suitability as a family pet. The dog who does his job with dour professionalism is at no disadvantage to the dog who is having a whale of a time. Dogs too shy to approach a judge, or dogs you wouldn't trust with children, are judged no differently from the agreeable beast who'll let tykes crawl all over him.

Many of the handlers will have judged trials themselves, and after the first scores are posted, they are quick to identify a judge's idiosyncrasies: "He wants a tight turn." Or, "He's quick to call the shed."

A judge can judge a dog of his own breeding, the dog he sold yesterday, the handler who gave him such an unfair score two weeks ago in Tucson. He can judge his own wife. In such circumstances, some wives and husbands choose to run "noncompetitively," but there's no rule about it. After all, as George Conboy liked to say, "It's just a bunch of black and white dogs chasing sheep around. No reason to get excited about it."

When Pip and I finished, we had a respectable score just out of the money. At the gate I thanked Lyle for passing on the judge's instructions, and the judge turned to say, "She didn't give you all of them."

At the handler's meeting that morning, before the first dog ran, the judge said he didn't want any dog sent before the handler reached the post, so I'd lost five points for that simple fault. Like George Conboy said, no reason to get excited about it.

4

First Rumblings

The United States Border Collie Club, founded in 1980 by horsewoman and dog obedience trainer Ethel Conrad, never had more than $3,000 in its treasury. Most of its 400 members were pet owners, with a strong minority of obedience competitors. Inside sheepdog culture, the USBCC was called "Ethel Conrad's club," and it drew its moral authority from Miss Ethel.

Whenever I turned onto the long gravel lane of Miss Ethel's Sunnybrook Farm, my dogs started grinning. They knew they had reached an outpost of Dog Heaven, one of those scarce places where everything makes dog sense. Over the years, I suppose, hundreds of Border Collies found their way to Sunnybrook, and when Miss Ethel finally passes through the Pearly Gates, she will find a wildly wagging, furry welcoming committee.

When younger, Ethel was a brilliant horsewoman. In Sunnybrook's hallway amongst leashes, collars, raingear, rubber boots, and hats is a photo of young Ethel jumping Black Watch. She remembers galloping at the fence, the biggest in the old Boston Garden, and the surge of Black Watch's body as the great horse left the ground, but then the photographer's flashgun blazed incandescent and she remembers nothing more until she woke up in the room "where they took the knocked out boxers."

"It took a lot to bring Black Watch down," Miss Ethel adds.

In those days she lived outside of West Point, and every morning, before she'd board the train into New York to her job, Ethel would ride Black Watch up the Hudson Palisades. It was pitch-black but she trusted the horse to find footing on the faint trails that wound along the precipices.

Three years later, when she and her new husband were settled at Sunnybrook, they learned Black Watch's owners were going to have the horse destroyed. The owners had sent the horse to a fool trainer, who'd ruined him. Bryan and Ethel offered to take Black Watch, to let him live out his life at Sunnybrook, but the owners had him put down. Ethel has not forgiven them.

Miss Ethel is quicker than most to speak her mind. At one trial, an overrated trainer had brought a coterie of novice disciples. To these beginners he loudly supposed that he was probably the best trainer in Virginia.

"That's a damn lie!" Miss Ethel snapped. The bogus trainer deflated like a balloon.

General Bryan Conrad was commander of cadets at West Point and Eisenhower's intelligence chief at the Normandy landings.

"Bryan was older than I was," Miss Ethel says, "and when I married him I hoped I'd have ten good years. As it turned out I had twenty-three."

The year Bryan died, Ethel's mother was dying too, and Ethel's days were between nursing home and hospital. And when she came home to her dark, lonely house so far out in the country, there was nobody to greet her except her Border Collies Spike and Arrow.

♠♠♠

Sometime before World War II, the American Kennel Club put the Border Collie into its "Miscellaneous Dogs" category, their junk room for dogs not good enough for full AKC recognition. How this classification came about is mysterious, and AKC archives are scant and unenlightening. Perhaps someone wanted to compete with Border Collies in the brand-new obedience competitions. Certainly none of the sheepdog people sought closer ties with the AKC. When Arthur Allen put on a sheepdog demonstration for the AKC in 1946, police cordoned off Madison Avenue so Arthur and his dogs could work sheep. AKC bigwigs smilingly agreed with Arthur that the Border Collie should never be exhibited in the show ring.

As an AKC Miscellaneous Dog, a Border Collie could compete in AKC obedience after it received an "Indefinite Listing Privilege" (ILP) number. Since the AKC knew nothing (and cared less) about the breed, they turned to the United States Border Collie Club (Miss Ethel) when anyone submitted an ILP application whose accompanying photograph didn't "look like" a Border Collie. There's a brisk turnover of junior AKC employees, and Ethel had to explain to each new staffer that a Border Collie "isn't what it looks like: it's what it does."

Ethel's club sponsored the first sheepdog clinic, the first judging clinic, the first advanced handler's clinic, and the Blue Ridge Trial, the first large sheepdog trial on the east coast.

But the American Kennel Club was the United States Border Collie Club's true *raison d'être*: we feared that one day the AKC would try to "recognize" the Border Collie, and keeping our dog out of their clutches was the USBCC's reason for existence.

For many years, that wasn't hard. The American Kennel Club admired dog aristocrats, whose unbroken lineages disappeared in the mists of fantasy. Dog breed clubs who yearned, nay prayed, for AKC recognition for *their* breed so their dogs could strut into the show ring with the proud Irish wolfhound, the spunky fox terrier, the fearless English bulldog, approached the AKC throne on their knees.

These supplicants had to convince the AKC that their breed's bloodlines were pure going back yea many generations, that the breed club represented the important breeders, and that the breed club had been in existence since kingdom come.

What followed was a lengthy courtship while AKC officials inspected the hopeful breed's studbooks (the record of generations of dog begats — the Burke's Peerage of a breed) and determined not only "are these our sort of dogs" but more important "are these our sort of people." It was not likely an impoverished minority group could have had their mongrels "recognized" by the American Kennel Club.

At the happy consummation, the breed club would turn over its studbooks to the AKC and were freed of the duty of registering begats and cashing begats checks. In return, their dogs could compete at AKC shows and their owners could mingle with the gentlefolk.

(When I explained dog registration to my friend Jake, Jake said, "I think I get it. I send you money, and you send me a piece of paper." After further consideration, Jake asked, "How do I get into that racket?")

Fortunately for the Border Collie, there were no Border Collie conformation shows, so there was no constituency of owners who'd shown their dog in dog shows and yearned to show in the Big Time. And since there were three Border Collie registries, if one registry lost its wits and sought AKC recognition, the other two would keep right on cashing begats checks. Finally, the Border Collie wasn't purebred by AKC standards. In the UK and the United States, if your unregistered dog competed successfully in sheepdog trials it could be registered on merit.("Yo! Molly, Guess what! Shep is a Border Collie after all!")

And such registrations did occur—of the Border Collie champions listed in E. B. Carpenter's National Champions of Britain and Ireland, twenty percent have a Register on Merit (ROM) parent, grandparent, or great-grandparent. To the AKC, Registry on Merit is mobocracy: might as well let the charlady mate with the king.

Miss Ethel's Club sought what she liked to call "cordial relations" with the AKC while keeping them at arm's length.

In June of 1991, the weekend after the Seclusival sheepdog trial, Ethel and I (as a USBCC director and Ethel's friend) were at the Oatlands sheepdog trial being cordial to Robert McKowen, AKC Vice President in charge of performance events. The AKC had recently begun its "Herding" program. It was (and is) a Mickey Mouse program and we didn't want Border Collies competing (few parents deliberately send their children to the worst schools), but we thought that anything that involves people with their dogs is good for the dogs, and if people want to try their shelties and beardies and Lassie collies on sheep, have at it and God Bless.

Something more sinister was in the wind. At a New England trial, McKowen told handlers the Border Collie was a fine candidate for AKC recognition, and maybe some of you fine people might want to become the AKC breed club? And for that matter, why was McKowen at Oatlands? I'd never seen an AKC official at a sheepdog trial. Why now?

Mr. McKowen ("Call me Bob") was a big, avuncular man with a W. C. Fields nose. He listened politely, but his eyes were bored.

When necessary, Miss Ethel can be dazzlingly charming and was that day. When she told him there was no constituency for AKC recognition, Mr. McKowen asked how many dogs our registries registered every year. He asked when the Kennel Club in the UK had recognized the Border Collie. He said that in the dogs at Oatlands he could see enough similarity to establish a breed standard.

I said there were twenty-three-pound Border Collies and seventy-five-pound pound Border Collies—which would he discriminate against? I said that you can't breed for performance and appearance; you cannot serve two masters.

Mr. McKowen said he had bred German shorthaired pointers, one of which had become a "dual champion" — show and field. We said that was interesting, but hunting dogs weren't sheepdogs. We said that British and Australian show Border Collies were completely useless on sheep. He invited Ethel and me to an AKC Herding Clinic later in the month, and we said, of course we'd come, we'd help if we could.

We ran out of things to say. Mr. McKowen enthusiastically described an AKC scheme for inserting microchips into field trial dogs to detect and prevent dogs that were already champions from running in trials.

I asked, "Why don't you just let them run?"

The Oatlands course was too tough for Old Pip. Gael's sheep were afraid of the great crowd behind the handler's post and she couldn't push them around. Neither could most of the other dogs that morning.

At noon, I had my fifteen minutes of fame signing copies of *Eminent Dogs, Dangerous Men* in the Oatlands coach house.

Signing books at sheepdog trials is probably a dumb idea. Unlike the dreaminess that drives minor book celebrity, running a dog is pure contact and focus.

When the last dog ran and the spectators went home, we all turned our dogs loose and they streaked around, low and fast in the twilight. It was Saturday night and Oatlands fronts US 15, a busy, two-lane highway. Pip and Gael and I sat outside my tent while daylight leaked out of the world. It was humid, and a zillion fireflies blinked hopeful messages. An unbroken stream of cars passed, but we three animals were invisible to drivers whose vision was reduced to two lanes and the glare in their mirrors. Sometimes car stereos were so loud cars shook and thumped. Dothump, dothump dothump. Pip sighed and rested his silky head on my knee. We humans like to think of ourselves as the rational animals.

Not two weeks later, the AKC moved to swallow the Australian shepherd.

Since it became a breed, the Australian shepherd had been registered by the Australian Shepherd Club of America (ASCA). ASCA holds sheep and cattle trials for their dogs, as well as conformation shows. Most Australian shepherd owners had no interest in the American Kennel Club, and in 1985 they had voted against seeking AKC recognition 699 to 307. ASCA wouldn't solicit the AKC on hands and knees, nor would they hand over their stud books, and they were perfectly happy to cash the begats checks themselves.

Without telling anyone (including the Aussie people) or the AKC's own membership, who never discussed or voted on these new procedures, AKC directors had changed their rules to (as they explained in a press release) "make the application and approval process more efficient and expeditious for breeds meeting AKC requirements."

In a nutshell, the new rules meant that if AKC staffers could locate (or create) a handful of breed owners who wished to join the AKC, the AKC could anoint them the dog's breed club, accept their application, and recognize any dog it wanted to. They'd forget about the studbooks — those precious repositories of begats — and simply take the word of those wishing to register their dogs with the AKC that their dogs were purebred Australian shepherds, and that if registrants claimed their dog was sired by decades of ASCA champions, by God, so it was.

For the American Kennel Club, this was somewhat like the Pope declaring that the Virgin Mary may not have been so virginal after all, maybe she'd been fooling around. The Aussie people were raging, phoning their lawyers, slamming the barn door shut on an all-too-empty barn.

And we discovered that a few obedience/agility handlers — people unknown to the sheepdog community — had formed a Border Collie breed club (The Border Collie Society of America) and were talking to the AKC.

I phoned an officer of the new club. She seemed like a nice woman who showed golden retrievers and wanted to show her Border Collies too. That dog shows would extinguish the Border Collies she loved didn't seem to worry her.

We were desperate to learn what was going on. You may think the AKC, a not-for-profit dog club, would be perfectly transparent, that its thinking and decisions would be open to public debate. Think instead of Byzantium, the College of Cardinals, a Chinese tong, the Supreme Soviet. As George Sangster wrote in 1986, "The Kennel Club is not an organization, it's a garrison. The world's largest secret society. The world's most impenetrable fortress. A stronghold, not a governing body. A psychosis, not a philosophy. It's in the twentieth century but not of it. A body completely surrounded by suspicion, a walled society, a kingdom of fear, an island in a sea of mistrust, a monument to paranoia." Couldn't have said it better myself.

Which influential breeder was cozy with which staffer, and what the staff was telling AKC Directors, was reported nowhere. Large circulation dog magazines depended on show dog advertising and reported on the AKC only when they couldn't avoid it. The mainstream press is uninterested in dogs. Dogs are kitsch, corny, sentimental — beneath a real reporter's notice.

The AKC had misled ASCA officials about its intentions and concealed the splinter group they planned to anoint as the AKC Australian Shepherd breed club. Although Bob McKowen disavowed AKC intentions of doing to the Border Collie what they had just done to the Australian shepherd, we can be excused for skepticism. When we asked for a letter from the AKC Board saying, in plain language, that the AKC had no interest in changing the status of the Border Collie, the silence was deafening.

The Border Collie community inundated the AKC with protests. They so jammed switchboards at 51 Madison Avenue that anyone who told the AKC operators he was calling about "Border Collies" got put on indefinite hold.

We feared our objections would be ineffectual, that at the monthly AKC Directors meeting some staffer would murmur that they'd gotten a few protests about the Border Collie and maybe wave a letter or two.

We had been thrown into a philosophical briar patch. What is a "dog breed," and who has legitimate interests in it? Do those who know the breed best have any authority over its future, or can any group do any damn thing it pleases with any dog breed? If, for instance, one group's breeding practices assured the genetic deterioration of very many individual dogs with consequent dog suffering and costs to the dogs' human owners, is there no way to stop it?

Breeding Border Collies for dog shows would inevitably produce a dog which was no longer a Border Collie. AKC recognition would—at best—create two breeds with the same name and—at worst, given the AKC's influence on dog-ignorant puppy buyers—AKC show dog breeding might supplant the useful Border Collie with a useless one.

We didn't know how quickly the AKC was moving, whether their next Director's meeting would recognize the Border Collie. We had to put a hitch in their gitalong, so we filed a trademark application for the name "Border Collie." We knew the application wasn't watertight (it was ultimately rejected because the name was already in common use), and we didn't have the money to hire trademark attorneys, but we thought that if the AKC acted right away our pending application would buy us time.

We asked people to write or call AKC Directors. We placed ads in obedience and sheepdog magazines. Signers of the ads included every National Finals winner, the top obedience people, and a few prominent citizens whose names, we hoped, would worry the AKC. We opened the Border Collie Defense Fund. One woman mailed us the first trial prize check she ever won. Another sent ten dollars with the plea, "Don't let them do to your breed what they've done to mine."

Sympathizers warned us that dog shows had ruined the German shepherd, Bedlington terrier, rough collie (the Lassie collie), the cocker spaniel, Doberman pinscher, Rottweiler, and Akita. To this oft-heard criticism, the AKC replies, "Registries don't ruin breeds, breeders ruin breeds."

♠♠♠

In June, as we'd promised McKowen, Ethel Conrad and I attended the AKC Herding Clinic.

Frying Pan Park was the historical farm outside Herndon, Virginia, where we'd been in March for a not-untypical sheepdog trial drenched with icy rain and the plowed ground so deep in mud some of the dogs couldn't force the sheep off the gravel road. My Harry (who hates to get his feet dirty) hopped from one patch of dry ground to another like a man in a business suit hopping puddles. The AKC would use a much smaller field and a long metal building which was three-quarters horse arena and one-quarter meeting rooms. We were there to get a commitment from the AKC to leave the Border Collie alone.

We'd interested Martha Sherrill of the *Washington Post* in our dispute, and Ms. Sherrill thought she and a photographer might come to the clinic.

Since we couldn't affect AKC decision-makers (indeed, we didn't know who they actually were) and we couldn't afford a legal fight, telling our story loud and clear was our best option.

When I arrived, Miss Ethel could scarcely contain herself, "Donald!" she said. "You are not going to believe this!"

"What?"

"Wait'll you see the sheepdogs!"

"Why?"

"Just wait and see."

Bob McKowen and Roberta Campbell, head of the AKC Herding program, invited us to dinner, and as Mrs. Conrad thought it rude to discuss controversial matters at their table, we made small talk and never brought up what was foremost in our minds. They did invite us to become AKC herding judges, and Mrs. Conrad explained that we'd be happy to help the AKC with their herding program, but we probably couldn't become judges until "this business with the Border Collie is settled."

The food — Holiday Inn? Ramada? — wasn't good, but Ms. Campbell and Mr. McKowen were agreeable hosts.

The next morning I met the sheepdogs Mrs. Conrad had been chuckling about. They were in dog crates stacked one atop another or on long retractable leashes, and when they came too near one other their owners reeled them in like fish. These dogs were classified as members of the AKC's "Herding Group" at AKC dog shows: Shelties, Bouviers, Pulis, German shepherds, rough collies, Australian cattledogs, a Corgi or two. They were handsomely groomed and their coats shone and they barked. They barked from their metal kennels and whenever another dog came near. One dog barked so continuously Mr. McKowen's audience couldn't hear what he was saying and he demanded that somebody find the dog's owner to shut it up. Owner arrived breathless, said she was sorry but she couldn't stop him from barking, attached the dog to her retractable leash and took him outdoors.

Mr. McKowen announced that the AKC was revising herding rules and that those people invited to contribute to the discussion had been notified, so no other opinions were required. There were questions about the new rules, to which

Mr. McKowen replied that after the AKC made its determinations, the rules would be available.

He and Ms. Campbell were brusque, completely unlike our affable hosts of the evening before. They treated the clinic participants as unruly, quarrelsome children and, to my astonishment, nobody took offense. Their arrogance might have earned them a bloody nose at a sheepdog meeting.

Miss Ethel's sheep were to be brought into the metal building three at a time so a Herding Dog could herd them around a simplified course. The AKC's most expert herders and their dogs had been flown in from all over the country to help, to put out sheep, to set up the course, to judge and scorekeep. Ethel whispered in my ear, "Do you have any idea how much this must have cost them?"

I guessed, "Ten, fifteen thousand?"

Miss Ethel smiled, "At least that," she said. (At the time, a *major* national sheepdog trial might cost $3,000).

These experts and their expert dogs couldn't bring the sheep onto the course, so the experts picked up the sheep and carried them.

"If they hurt my sheep, they'll pay for them," Ethel announced grimly.

An acquaintance of Ethel's took a seat beside us. Ethel introduced him as somebody who was "in on" AKC's secrets.

When we asked about the Border Collie, this AKC expert pronounced, "You're already in. The Directors voted you in at the last meeting. They're just waiting to recognize the new club as the breed club before they go public."

He spoke without hesitation: 2 and 2 equals 4.

In the horse arena, some very nice dogs were being asked to do something with sheep, but the dogs didn't know what and their owners didn't either. The dogs would run out at their owner's command but when they came behind the sheep they promptly lost interest. Some went to ringside to schmooze the spectators, some sniffed the sawdust. A few were afraid the sheep might hurt them.

Might as well ask a turtle to fly. When the paid experts' dogs went to retrieve the sheep, they'd come around briskly like the well trained beasts they were, and then they too would lose interest. The paid experts ran beside their dogs, clapping and stirring them up. The dogs would bounce around, excited, some even looking at the sheep, and after they got enough experts out there with their expert dogs the sheep would come off the course. No single dog actually *ran* the course.

I found Vice President McKowen and said, "It's time we talked business. Does the AKC intend to change the status of the Border Collie?"

He said, "I don't personally know of such plans."

I said, "I didn't ask that. I asked if the AKC intends to recognize the Border Collie."

"Personally, I don't know that we do."

I asked him to call New York. I said we needed an unequivocal statement from the American Kennel Club, something in writing. I said *The Washington Post* was here and only an assurance from the AKC would prevent bad publicity.

Mr. McKowen did not pick up the telephone.

That afternoon, Ethel and I did a demonstration so the AKC experts could see what sheepwork looked like.

Our demo was to be in a smallish field, probably a hundred fifty foot outrun. AKC experts were laying out the course with a tape measure.

Since the course was short and narrow, there wasn't room for the sheep to get through the first obstacle and make their turn toward the second, and I said as much to the expert setting up the course. I am bad with names and don't recall his but do remember he had a Puli—a black ropy-haired Hungarian dog that had no more interest in sheep than my boot does.

I told the expert the sheep would have difficulty and he said, "We measured it with the tape."

I said, "Look, suppose I'm a sheep" and trotted toward the obstacle. I trotted through and wrenched around trying to hit the second one. I said, "Look, it's no big deal, just move the panels back ten feet. This is just a demonstration."

The expert drew himself up to his full height and said, "I've run in three hundred sheepdog trials."

My jaw dropped. I gaped. "But, bu . . . why don't I know you?"

It was a simple course, docile sheep, and both Gael and Miss Ethel's Tess did good work. Afterwards, while Gael was cooling off in the water tub, the *Post* photographer took her picture.

Later that afternoon, the AKC herding people tried to put on a "tending" exhibition. They'd flown a woman and her German shepherd in from California to demonstrate "tending." Mrs. Conrad and I relaxed in lawn chairs on the verge of the field where the demonstration was to take place.

Unfortunately, when the expert saw the sheep she broke down in tears. "My dog isn't ready for this," she sobbed. So they took the sheep away and people became ersatz sheep. These people ambled around aimlessly, pretending to graze. Some said "Baaaaa." An ersatz shepherd stood where a tending shepherd might stand while an ersatz tending dog roved up and down to prevent the ersatz sheep from straying onto the ersatz autobahn.

♠♠♠

We did not expect the *Washington Post* to treat the AKC gently and they did not.

> The German Shepherd looks lost. Way out in the middle of a big field—Frying Pan Park in Herndon—he doesn't know where to look. At his owner? The audience? Certainly not . . . that stupid dirt cloud of fleece.
>
> His owner urges him to herd. What? Commands are shouted, "Fetch" and "Drive!" The German shepherd, looking at the five sheep, has an occasional flash of insight but nothing comes of it . . .
>
> Other dogs try and fail: the Belgian Sheepdog, the Old English sheepdog. This is a "herding clinic" put on by the American Kennel Club (AKC) for its fourteen "herding" breeds. Two Belgian tervurens stand dumbstruck. The Shetland sheepdog runs about nervously. The Hungarian Puli—which has a dull black coat of dreadlocks—seems distracted by his own fur. ("The Border Collie: A Breed Apart," *Washington Post*, B2, July 16, 1991)

Ms. Sherrill saw what any child could: that the AKC "herding breeds" and their owners were clueless. She wondered why the AKC, an organization that knew nothing about the Border Collie, should want to take it over. When she telephoned Ronnie Delay, head of the would-be AKC breed club, Ms. DeLay said, "I have a family. I don't have hours and hours a day to put into teaching my dog to herd like that. The AKC trials are more my speed."

♣♣♣

Harry was Gael's son, fourteen months old: a powerful, shortcoated, big-headed, ugly Border Collie who looked like a cross between a black and tan coonhound and a malamute. When Harry worked, he laid his short sharp ears flat on top of his head. Harry's scraggly white ruff reminded me of Charlie Chaplin's tawdry elegance.

Life ran in torrents through Harry's powerful body, and Harry didn't always look where he was going. He whammed into things: humans in the way, other dogs; once, full-tilt, a tree.

Harry was a "natural dog." A "natural" sheepdog makes a wide outrun without human help, walks up nicely on his sheep, and brings them straight to a handler's feet. Harry's sire was Tommy Wilson's Roy, who'd three times starred on the David Letterman show, most memorably when he closed Letterman's tenth-anniversary show by moving sheep through Rockefeller Center, through the lobby, out the doors into a waiting Checker cab.

Sheepdog people remember Roy's brilliant shed at the 1992 National Handler's Finals in Sheridan, Wyoming. Roy and Tommy faced fifteen sheep in a chalkmarked hundred-foot ring, five of the sheep wearing broad red ribbons round their necks. Without touching them Tommy and Roy had to separate off the ribboned sheep (only the ribboned sheep) and take them away. The judge would deduct points for false starts, for a dog refusing to come between the sheep, and for any sheep that left the ring.

Roy glanced at a sheep and that sheep said "Anything you say, boss," and separated itself. Roy glanced at a second sheep. It separated itself. Presently, there were two groups of sheep in that small ring: ten unribboned, five ribboned. Tommy and Roy took the five ribboned sheep and popped them into the pen. I wasn't the only one with tears in my eyes. At my elbow a tough old ranch woman muttered, "We ask so much of them. And the damn fool dogs go out and do it!"

Roy's son, Harry, had so much potential I worried I wasn't a good enough trainer to educate him. What Harry did naturally, he did so well and so easily, he resented my commands and when I scolded him about it, often as not, he'd leave and go home.

He yearned to be a stud dog and would have looked swell in a black leather jacket.

♠♠♠

I'd written Louis Auslander, Chairman of the AKC Board of Directors, arguing that AKC recognition would harm the

Border Collie and we'd resist such recognition by any means possible.

After the debacle at Frying Pan Park but before the *Washington Post* article came out, Mr. Auslander replied that I had apparently believed a "little lie"—to wit, that the AKC was a dog show organization with scant interest in performance and that anyway they weren't scared of us. He was willing to meet.

Miss Ethel took Auslander up on his offer, and on a bright August morning we met our attorney, Ray Mundy, in the cavernous lobby of The American Wool Building, 51 Madison Avenue: the AKC's New York Headquarters.

"What can we expect in there?" I asked.

"They've been unchallenged for so many years they are indifferent to everything outside their own doors," said Ray Mundy. "What matters is their internal politics. One faction is trying to oust Auslander. Expect them to be very angry with you. They didn't like that *Washington Post* article, count on it."

The reception area of the American Kennel Club was decorated like an English club: fine wood paneling and some first-rate dog art. One nineteenth-century painting depicted a hunting party in the highlands. The sports were asking a craggy kilted shepherd where the game was, while at their feet, unnoticed, their fine sporting dogs encountered the shepherd's collies, lips curled, baring white teeth.

AKC President Bob Maxwell welcomed and ushered us into the Board Room—long walnut table, heavy chairs, water pitchers and glasses, no windows. The AKC staffers filed in, looking glum. Vice president this, and vice president that and of course you know Mr. McKowen. We introduced Ray Mundy, our attorney, and a staffer shot out to fetch the AKC corporate counsel.

Louis Auslander was a lanky, long-faced gentleman who shook hands perfunctorily, took his seat at the head of the long table, and promptly lit into us about the *Washington Post* article. Hearing his cue, Bob McKowen got red in the face and joined the attack. Auslander was outraged at the letters and phone calls from Border Collie people but more outraged at the *Washington Post*.

I said that if they'd been responsive, if they'd said they had no intention of recognizing the Border Collie, there would have been no letters, no bad press.

Auslander replied, "We can stand the heat."

Vice President John Mandeville said that many breeds had show and working lines, but they were still one breed. We asked how naive puppy buyers would know whether their pups would work stock or not. Bob McKowen said some show champions were champion field trial dogs too, and we said that wasn't the case with sheepdogs. Miss Ethel quoted a British show judge deploring the loss of working abilities in show Border Collies.

(In a 1990 article in *Gundog Magazine*, Ken Roebuck commented that show and hunting springer spaniels

are about as much alike as chalk and cheese. . . . A springer spaniel is a springer spaniel insofar as the American (and British) Kennel Clubs are concerned whether they be from hunting or show lines. . . . Some breeders of show springers possess the integrity to tell inquirers who are searching for a potential gun dog that theirs are not the type they should consider. . . . There are many others, however, who possess no such scruples, who will glibly tell anyone "Sure our pups will hunt. They're springers, you know. They all do." . . . Of the five (show springers) we took in (for training) over the past two years, only one made the grade. Two lacked all desire to work, but worst of all, the remaining two had unstable temperaments. One of them would have bitten if given the opportunity. . . . And these were spaniels for heaven's sake. Dogs that since time immemorial have been recognized for their kind, gentle nature)

I told them that Border Collies were often registered on merit, that my bitch Gael had three such ROMs in her pedigree. From the looks I got, I might have been advocating kinky sex.

Auslander said that the Border Collie had been in the Miscellaneous Class for 28 years, that it was just "parked there" and that it "didn't do us (the AKC) any good."

I said that we'd back an increase in their fees.

President Maxwell said, "This is not about money."

Auslander told us that the AKC only registered 130 breeds while the Kennel Club (UK) registered 140 and the Continental Kennel Club (Europe) registered 400 breeds. Over time, he said, the AKC intended to register "every registerable breed." He said that the Border Collie was very low on their list of priorities, that our concerns were "paranoid"—that they had plenty on their plate with the Australian shepherd.

"That's another thing . . ." I began, but Miss Ethel kicked me in the shin.

Vice President Mandeville wondered what the harm was having two "strains" of Border Collie, working and non-working, that sort of thing went on all the time.

I asked their legal counsel, "How would you feel if I took an office across the street, called myself the AKC, and started registering dogs."

Their counsel said, "We'd take you to court."

I said, just so. I asked what would be wrong with having two "strains" of AKC? If any of them got my point they didn't say so.

Auslander said they were re-evaluating the Miscellaneous Class and I said since we knew the dog better than anyone we wanted to be consulted about any decisions involving the Border Collie.

Auslander said that would be okay.

After the meeting, when Bob Maxwell handed me my coat, he said, "You shouldn't say we do this just for the money. Do you write just for the money?"

I was dumbstruck.

♠♠♠

Dog collars hang on nails beside our kitchen door. The tiniest collars are for puppies; broad leather collars for the older dogs. No choke collars. Though I've been training working sheepdogs for years, I've only used a choke collar once. I haven't anything against them; I just don't use them. People think collars are for controlling a dog, but they aren't.

You can spot these people in the park being towed behind some great brute of a dog, anywhere the dog wants to go.

Steel leashes and collars don't give you control over your dog; your dog gives you control over your dog.

Collars mean more to humans than they mean to dogs. When sons-of-bitches abandon their aged, trusting family dog along our country road, they always remove its collar before they drive off.

At big trials, handlers sometimes remove the collars before they send their dog. With half a mile between man and dog, the collar's just a distraction. Once I timed a sheepdog taking commands — seventeen commands in fifteen seconds. No, the dog wasn't wearing a collar.

I put collars on our dogs when I take them to town and during hunting season. The collars have brass nameplates with my name and phone number so if the dogs get lost, kindly people can let me know where they are.

I put a collar on when I take a dog to the vet. Old Pip was in so much pain he couldn't lie down or drop into the sheep-dog's crouch.

In the vet's office, I said he should get into the cage where they'd put an IV into him and he did. When I lifted him onto the operating table, he smiled at the vet tech and licked her hand. "He likes girls," I said. "He always likes the girls."

The cancer had filled his body cavity and wasn't operable, so I stroked his silky head and murmured the words you say while the vet put the killing stuff in the IV. I carried him wrapped in a sheepskin out to the car. Anne and I buried him on the hill above the meadow where the sheep bed down and I train the young dogs. I hope Pip gives the young dogs a tip or two. I know he advises me. I hung his collar on the nail beside our kitchen door. The collars don't belong to any dog special. Any of our dogs can use them.

5

Silk

I'm a little teapot, short and stout.
Here is my tailpiece,
Here is my snout.
When you roll me over,
I will shout.
Please don't throw this good dog out!

Show breeding is so triumphant that most dog pound mutts are no more than three generations away from purebreds who competed in the show ring. Americans have accepted the dog show credo: "a dog is what it looks like."

What is a Labrador retriever? An English sheepdog? A bulldog? Don't we picture a squat black dog, a dog with hair over his eyes, a dog that looks like Winston Churchill? For most of us these names do not conjure images of a dog swimming to retrieve a dead bird, or driving sheep along a narrow road or facing a bloody, maddened bull.

Most people have no use for a pet which might savage their children or cannot be housebroken. None would knowingly select a puppy likely to go blind or lame in adulthood.

But when a puppy buyer seeks a purebred to live with for ten, twelve, sixteen years, his menu consists almost entirely of dogs bred for the show ring where neither temperament nor trainability nor soundness nor the ability to perform those tasks which are part of the dog's name ("retriever," "sheepdog," etc) are valued and may indeed be deliberately selected against.

The June 6th, 1874 *Field and Stream* reported an "exhibition of dogs without any attempt at judging their hunting qualities" held by the Illinois State Sportsman's Association. A scant thirty years after this first American dog show, no dog could be exhibited at the St Louis World's Fair unless it was purebred — i. e. from show dog breeding — and registered by the American Kennel Club. This was a stunning conceptual change. In the history of American domestic animals, only the disappearance of horses from our landscape after World War II is as dramatic.

♠♠♠

A few days after our meeting at 51 Madison Avenue, I wrote Chairman Auslander thanking him for his willingness to consult with us before changing the status of the Border Collie and offering to demonstrate Border Collie abilities for him or any of his directors. Chairman Auslander promptly wrote back that they'd do whatever they wanted and had no plans to consult us, thank you. He didn't reply to my offer to show his directors the dog we were all talking about.

I wasn't surprised. Before this fight started, I had viewed the AKC as a dumb but benign dog-friendly organization with an eccentric interest in dog shows. I expected our arguments about what was good for our breed to carry weight. Since it was demonstrably true that Border Collies were unequaled as stock dogs and just as true that stock dogs taken into the show ring were, in a few generations, useless, I thought that should settle the matter. I hadn't planned to mention that AKC show ring practices had reduced the rough collie, Old English sheepdog, Shetland sheepdog and bearded collie to uselessness—I had no wish to insult anyone's dog. I have no objection if you want a collie. I won't demur if you say that "he still has all his herding instinct. You can't keep him from rounding up the children." Love is notoriously blind. But I had hoped to convince the AKC dog people what every sheepman knows: If your livelihood's at stake, get a Border Collie.

The AKC was indifferent to our arguments, no matter how diplomatically couched. Most of the AKC people were friendly but they didn't seem to "get it," nor did they seem to care. Their language was peculiar, but I assumed we spoke different dog dialects. After all, when a sheepdog man says, "She bumped them at the top" or "He came in on the wrong sheep" structurally, this is not so different from the dog fancier's "He needs another major to finish."

♠♠♠

We Border Collie people thought we were dealing with a simple misapprehension: the AKC didn't know what our dogs were and that after we told them, they'd either register Border Collies without pushing them into the show ring or go away and leave our dogs alone. We had no idea they'd heard all our arguments a hundred years ago when the dog fancy first usurped the breeding of dogs. In dog fancy culture, our arguments were the tiresome whine of a crank so out of the mainstream he needn't be rebutted.

In 1904, W. A. Sargent wrote in *Field and Fancy*:

> In speaking of the brains of the Collie, or more particularly of the usefulness of Collies in driving and herding, one man says, "It is something that cannot be altered by breeding, nor lack of use for generation after generation or from any other cause whatsoever. It is as firmly seated as an instinct in the Collie's brain." Let me ask, just simply ask, why then, if you can successfully breed for heavy coats, long noses, prick ears and vice versa you cannot equally as well breed for brains?
>
> I am not a sorehead, have nothing against the judges or the other "wise ones," have never exhibited a Collie in my life, but have bred reared bought and sold hundreds of them . . . . I know a Collie that has won several good prizes, firsts, seconds and thirds, in good competition, that has as few brains as an elephant has feathers. I would not take him as a gift. Wouldn't give forty cents apiece for his get, and yet he wins. He wins when other Collies that are placed away below him would make him look like 'thirty cents' if judged on their merits. . . . The majority of Collies that are shown on the bench belong to gentlemen who are well able to have and sustain large kennels. They breed for show points, and to show points. So far as I know about any show in this country, the Collie is given no chance to show his mental powers, beyond looking bright, attentive and attractive. The majority do

nothing but simply exist to be exhibited for their good looks as ruled by the accepted standard or the judges. Displayed intelligence does not count in the shows; who then is going to the trouble to breed for it?

Although most of the early literature was written by and for dog show enthusiasts, objections like Mr. Sargent's were not uncommon. Judith Neville Lytton—Baroness Lytton— was the fiercest, most thorough critic of the British dog fancy. She saw nothing wrong with dog shows, but deplored their claims to breed antiquity. In *Toy Dogs and Their Ancestors* (1911) she writes that that "present standard and scale of points has apparently no foundation earlier than 1885 or 1887." She thought breed purity was a myth:

> Compiling a history of the Toy Spaniel breeds has been like unraveling a Chinese puzzle. The errors of translators and the abnormal amount of hypothesis to be sifted have made me feel at times like the poor princess who was given four sacks of feathers of hundreds of different birds and told to sort them into their proper species before midnight. . . . The Blenheim isn't a Blenheim, the King Charles isn't a King Charles, and the Pomeranian is not a Pomeranian at all.

Lytton also anticipated modern critics:

> The whole fabric of modern judging is utterly unsound, The Club judges (KC) are, moreover, bound by the Club regulations which prevent the exercise of any private judgement. . . . for instance a good sound dog, perfect in all points, will be

put back by practically any specialist judge for white on the chest, and a glaring cripple preferred to him, provided it has no white hairs.

Lytton called the dog fancy's toy dogs "Noseless Monstrosities" and described the Kennel Club's reaction to criticism. "There are many fanciers who deplore the ways of the dog fancy as much as I do, but if they speak up they are put into coventry and Good-bye to all hope of winning with their dogs."

So far as I can determine—and I've looked—no dog fancier ever rebutted Baroness Lytton. They ignored her.

Then, as now, the dog fancy preferred roundabout rebuttals.

Roundabout rebuttal #1: The show dog that outworks all the best working dogs is still with us. (Dog fanciers tell me about him all the time. It is regrettable this paragon is never seen at a trial or anywhere else he might be tested.) An example of this rebuttal was published in *Field and Fancy* in 1911:

> I believe the Collie of today, if properly trained, can and will make a perfectly good cattle and sheep dog. The best one I have seen was one belonging to Mr. Piggin, of Long Eaton, Derbyshire England, and owned champion Christopher for sire. Ormskirk Charlie he was called, and his intelligence was almost human. He beat, in his day, all the best dogs [*nota bene*] in England and in Wales, and he was besides a winner in good company on the show bench

Roundabout rebuttal #2: Addressing points the critic has not raised. In *A History and Description of Modern Dogs* (1894), Rawdon Lee admitted that the show collies were no longer useful but

> He is a different dog now, when well attended and cared for, than he was when he had less value. The churlishness and snappishness, which were prevailing features, appear to have almost entirely disappeared, and he does not rush out of the farmyard and seize the passing wayfarer by the calf of the leg, by the coattails or elsewhere as was once his habit. Constant association with his superiors has improved his disposition immensely, he has risen to the occasion and to his aristocratic surroundings.

Dog fanciers had "refined" the collie.

♠♠♠

The *New York Times* reported that Berkeley geneticist Jasper Rine was conducting experiments using a Border Collie. The *Times* didn't specify how Rine's dogs were being treated so I fired off an inquiry. He reassured me that the dogs lived with him and his fiancé and that resulting pups would live out their lives in good homes. He sent scientific articles the Dog Genome Initiative had already published and hoped we could get together when next he came East.

I invited him to the farm.

Rine pulled down our drive in mud season, early spring. The Director of the Human Genome Project and Dog Genome Initiative at Lawrence Berkeley Laboratory was a lean, courteous, intense man. As a teenager, Rine had worked on a dairy farm in upstate New York with a Border Collie and hadn't forgot how to do hard, grubby work. Together we'd load our '53 Dodge PowerWagon with hay. In the field, I'd toss hay off the back of the creeping machine while Rine slogged through the muck opening bales and scattering them for our hungry ewes.

He asked about Border Collie genetic "behaviors" (his word). I said they were quite complex, that some Border Collie strains ran wider on their outrun than others, that most were thunder shy and some were spooked by the slightest noise. That some strains seemed to be more powerful — the sheep respected them more — and that these type of dogs were often harder to flank (shift from side to side).

I told him what Tommy Wilson had observed of unintentional crossbreds in Scotland. They (the crossbred offspring) "have some of the bits but not all of them. Oh no. You can't train them. They're no use."

Jasper Rine told me the Genome Initiative was measuring the dog genome and planned to match individual behaviors to locations on that genome. Jasper's a better teacher than I am a pupil and though his explanations made perfect sense from his mouth, I couldn't have repeated them five minutes after.

I offered to connect him to working Border Collie people in California who would know how to evaluate his dogs, and he invited me to come to Berkeley and see the experiment first-hand.

After three days of winter chores, morning and evening, Jasper was mudspattered and unshaven when he carried his duffel out to the car; off to see Dick Darman, President Reagan's budget director.

"Don't you want to clean up, first?"

"Oh, no, no. I'll get respectable at the hotel."

And climbed into his rental car and roared up the hill.

♠♠♠

The AKC fell silent and so did we. "Maybe we've misjudged them," someone hoped. "Maybe they'll leave us alone."

I thought they were still digesting the Australian shepherd and didn't have gut room for us. There wasn't much more we could do. We made a contact list so we could move quickly when they came at us again, and in May I went sheepdog trialing.

♠♠♠

Chairman Auslander lost his battle and his job. As it happened, Miss Ethel knew the new Board Chairman, Jack Ward, slightly from the time many years ago, when she competed in AKC obedience. Miss Ethel wrote Ward her congratulations and repeated our offer to demonstrate what our dogs were and could do. She hoped that "our historic good relations with the AKC could be restored." Jack Ward took her up on her offer and they met at a demonstration Ethel was giving at Montpelier.

Jack Ward was more sympathetic than Auslander and promised that so long as he was AKC Chairman nothing would be done to change the status of the Border Collie. His promise was hedged; he refused to commit either the AKC or its next Chairman to leave us alone, so we were incompletely comforted. But clearly this was the best we were going to get. In exchange, Ward asked us to shut up and turn away any reporter's inquiries. Since such inquiries seemed unlikely, we were willing to give up our First Amendment rights in exchange for a truce. We had five thousand dollars in our Defense Fund. We thought we might yet need it. Life went on.

♠♠♠

One of our sheepdogs was dying. Silk had an osteosarcoma, an acorn-sized growth on her muzzle, too near her optic nerve to operate. Though the vets said Silk might live an additional three to eight months if we drove her to the University of North Carolina's vet school for chemotherapy, with an eleven-year-old, deaf sheepdog, it was hard to see the point. Silk felt fine; she was unaffected by her cancer. When it got bad, I'd bring a vet out to the farm so Silk could die in her home.

Gael slept on a corner of the couch, Silk behind her chair, Harry on his dog pad, Moosie hidden behind the dresser. For the dogs, living indoors was a trade-off: they lost the peace and quiet of the kennel, were caught up in every domestic emergency and learnt a range of household behaviors that didn't interest them much. In exchange, they were closer to humans and for reasons I didn't understand, that's their preference.

As a pup, Silk showed great promise. Her sire, Jack Knox's Craig, was the best working sheepdog in North America, and her dam, Jed, could make my heart stand still. I remember a time on a farm outside Winchester, when a goofy young ewe jumped a fence and got in among a herd of wild cattle. That sheep knew that the cattle would attack any dog that came near. Despite their charges, Jed peeled that ewe away in thirty seconds and pushed her right into Jack Knox's arms.

Many of the best Border Collies are homely. It is considered mildly insulting to tell a sheepdog man how "pretty" his dog is. With her black and white mottled coat and blue and brown eyes, Silk was pretty. Gael, her rival, was half Silk's size, black and tan, ratty looking, and Gael yearned with every ounce of her bitchy little heart to be Queen of the Pack—a wish she'd have gratified soon enough. Meantime, whenever Gael pestered Silk too much, Silk put her in the closet. With nary a snarl or growl, using pure moral authority, Silk marched forward and Gael disappeared behind the coats and shoes.

Silk's hearing loss was progressive. Border Collies are taught their trade from eight months to 30 months of age and Silk, who picked up the basic stuff easily, couldn't go on to the next step, in which the collie drives the sheep away from you by relying on whistles. I am ashamed how long I failed to understand that the reason she wasn't doing as I asked was because she couldn't hear my asking. I finally taught Silk hand signals, but because my dogs often worked half a mile away and sometimes out of sight, Silk was almost useless.

Silk did like to load sheep. She was both gentle and implacable and the sheep clambered up the loading chute into the truck without protest. Every week last April, we loaded lambs for market, and after I backed the truck up to the chute, I'd let Silk out of the house. By the time I got back to the truck, she'd have 22 lambs loaded and waiting to go.

As the least able dog, she was the most neglected. Although I often took a dog to town, it was never Silk. When a neighbor phoned — he had cows out, could I bring my dogs and help? — I didn't bring Silk. In the afternoon when I took a dog out for refresher training — Border Collies get rusty just like we do — it was never Silk. When I went to sheepdog trials, Silk's hated rival Gael went with me; Silk stayed home.

After we learned Silk was dying, Anne and I decided to do something for her so we wouldn't feel so guilty about neglecting her. But what can you do for a dying dog? Silk wouldn't care that her name's in this book; she was indifferent to bright, noisy pet toys. If Silk could read a menu, she'd order deer liver and lamb kidneys, *au natural*, but afterwards, she'd go out in the yard and eat grass. Many pet owners think their dogs are chowhounds, mistaking dog courtesy ("these humans are excited about this stuff, maybe I should show some enthusiasm") for unbridled appetite. Silk couldn't care less.

Dogs are symbolists, exquisitely attuned to correct gesture, assuaged by custom and ritual. Silk liked it when Anne took her on solo walks because, in a four dog pack, it was a privilege to be walked alone. I took her out for training. I'd stand in the middle of our flock and signal her: go left, go right. I'd show her the bottom of my palms: lie down. I'd signal her to walk up on her sheep; when they broke, she'd cover them. The old, deaf, dying dog ran around the sheep until her tongue hung out.

Afterwards, she'd come up to me for a pat and she was quite pleased with herself: "Aren't I such a good sheepdog?" And when she went back into the house, she promptly put Gael in the closet

All that we can give a dog that the dog will value is our time. Of which I had more than Silk did.

6

Two Dog Shows

Charles Dickens wove dogs into many of his novels. In the August 1862 issue of his journal All the Year Round, *Dickens wrote about "Two Dog Shows."*

It has been said that every individual member of the human race bears in his outward form a resemblance to some animal; and I really believe that (you, the reader, and I, the writer of these words, excepted) this is very generally the case. Everybody surely can with ease point out among his friends some who resemble owls, hawks, giraffes, kangaroos, terriers, goats, monkeys. Do we not all know people who are like sheep, pigs, cats, or parrots; the last being, especially in military neighbourhoods, a very common type indeed? Let any one pay a visit to the Zoological Gardens with this theory of resemblances in his mind and see how continually he will be reminded of his friends. Among the aviaries, before the dens in the monkey house, and even in the serpent department, he will find himself *en pays de connaissance* at every turn. But what is more remarkable is that there is one single tribe of animals, and that the most mixed up with man of all, whose different members recall to us constantly different types of humanity. It is impossible to see a large collection of dogs together, without being continually reminded of the Countenances of people I have met; of their countenances and of their ways.

In that great canine competition which drew crowds some week or two ago, to Islington, there were furnished many wonderful opportunities for moralizing on humanity. It was difficult to keep the fancy within bounds. With regard to the prize dogs for instance (to plunge into the subject at once), was there not something of the quiet triumph of human success about their aspect? Was there not something of human malice and disappointment about the look of the unsuccessful competitors? Was there not a tendency in these last to turn their backs upon the winners, and to assume an indifference which they did not feel? There was a certain prize retriever, and a more beautiful animal never wagged tail. To see that creature sitting up and looking with an air of surprise towards that direction in which some other (and probably unsuccessful) dogs were making an immense noise with discontented growls and barkings—to see his calm expression and utter want of sympathy—was a great sight, and the curled-up disgust of the other retriever, who had failed, and whose position was next to that of the prize dog was an even greater sight. On the whole the winning dogs carried their honours with calmness, with the exception of the prize King Charles Spaniel whose nose was a thought arrogant, sustained their triumph with modesty and forbearance. It is not difficult to occupy the first place becomingly. The winners of such prizes can afford to be quiet and unassuming. But to feel that you can retrieve better than the prize retriever, that you can hang onto a bull's nose better than the prize bulldog, that you can make yourself generally disagreeable better than the prize lap-dog, is a worrying thought for the second class competitor, and is apt to make him curl himself up and snap and render himself in a variety of ways hugely unpopular. For it is to be supposed that the prizes in this same dog competition were accorded more to perfection of canine form than to intellectual merit, there being no opportunity to form an estimate of a pointer's pointing, a retriever's retrieving, a bulldog's bullying or an Italian Greyhound's aggravating in the agricultural hall at Islington. To take the owner's word for the abilities of each animal would be of course out of the question.

The beauty of one dog, the ugliness of another, and of all the utmost development of the individual peculiarities of the species to which they belonged, would seem to be the causes operating with the judges. Prizes are to be won by size, by depth of chest, by clean finish of limb, and symmetry of points as in the case of the setter, the retriever, the greyhound, the pointer. Meanwhile to be bandy, blear-eyed, pink-nosed, blotchy, under hung, and utterly disreputable, is the bulldog's proudest boast. The bloodhound's skin should hang in ghastly folds about his throat and jaws with a dewlap like a bull. The King Charles spaniel wears a fringe upon his legs like a sailor's trousers, and has a nose turned up so abruptly that you could hang your hat upon it if it were not so desperately short. The prize terrier wins because he weighs two pounds and three-quarters, and the boarhound wins because he would (to look at him) turn the balance with a Shetland pony in the other scale. Truly the qualifications of dogs are numerous, and very various their claims on our admiration. We give a medal to a Cuban hound for tearing down a fugitive slave, and to an Italian greyhound for wearing a paletot and trembling from head to foot (I saw him) when a fly enters his cage.

It is a great comfort not to understand a subject. When I enter a friend's garden and sniff and stare about me, how I enjoy the perfume and the colours of his flowers, what childish days they awaken, and how happy I feel. The Scotch gardener has another Scotch gardener, and friend, to see him, and together they go the rounds of the beds. They only think whether this is a good specimen, whether that is "doubled," or the other equal to the example exhibited by Mr. Dibble at the recent rose-show. How contentedly my friend Corker refreshes himself with that claret a connoisseur would pronounce undrinkable; how happily he sits behind that carriage horse with the disgraced knees.

Now, if I had understood dogs, what sort of a visit would mine have been to the show? I should, like the Scotch gardeners, have gone about comparing "specimens," and carping as I heard many wiseacres do, at the decisions of the judges. What time should I have had for speculating as to the respective sensations of the winning and losing competitors? What opportunities for twisting a look of disappointment out of the features of one dog and a look of triumph out of another? Should I, again, if I had understood dogs, derived the pleasure I did derive from discovering that the prize terrier, which was about the size of a rat, was the property of an immensely big man, and so instantly darting off to the conclusion that all the little dogs belonged to big men and all the big dogs to small men. This exquisite theory, which no amount of examples to the contrary will ever shake me out of, would never have dawned upon me had I been a dog fancier. On the contrary, I should have journeyed about among those delightful animals entirely blind to their more wonderful qualities. I should have talked about "a man I know who had a pointer who could lick any dog in the place into fits"; or I might even have gone to the length of remarking that if "Manger of Stayleybridge had sent that bitch of his, she would have taken the shine out of any of 'em." Apropos of the foxhounds, I should have related extraordinary performances in a run with the Quorn hounds after a certain vixen-fox, which the whipper-in said was a dog-fox the moment it broke away, but which "I knew was a vixen-fox, and so it turned out." Before the pointers, I should have discoursed again of shooting and would, perhaps, have gone the length of saying, "Ah, many's the day's shooting I've had with that very dog, for I always go over in September to Sir Thomas's, and a capital cover it is. Here, Ponto, Ponto." Ponto would, perhaps, have failed to recognize me, and, perhaps, would have rewarded my caresses with an attempted snap, but still I should have gone on in the same way, and even the old spotted spaniel of the story book illustrations, a spotted spaniel would have been to me, and nothing more.

That Clumber spaniel is unquestionably the old original dog of one's childhood. One's first acquaintance with the canine species was made through the agency of the coloured story-book, and it was one of those spaniels which figured on the page. His name was Dash, and the tan spots were dabbed on in water-colour so boldly that they bulged in many places over the outline of the animal's form, and covered portions of the background (not to say miles of remote prospect) to which they were but indifferently appropriate. But it is not entirely owing to ancient associations that these dogs are so attractive; they are really most beautiful and rare animals. A dog is a great bore; he howls in the night; he is tiresome to feed; he wants to go out when you have somewhere to go on business and cannot take him, on which occasions to see him with his head on one side looking after you as you shut him up is enough to break your heart; he is disliked by your friends whose carpets he impairs and whose cat he frightens; he is liable to be stolen, and to catch distempers and other diseases, in short — he is altogether a heavy handful, but still if anyone were to offer me one of those real old fashioned spaniels, I hardly think I could refuse.

And if these Clumber spaniels are full of old associations, so also is a curly-haired coloured retriever. To see one of these dogs is to think of some old squire in the country as he makes his rounds about his gardens and farms, armed with a walking-staff with a spud at the end of it, who is sure to have a superannuated retriever of this sort at his heels. A good dog he has been in his time but he is now past his work, and is admitted as a privileged animal to the drawing rooms, is fed with bits of biscuit after dinner and listens to all the squire's directions which he gives to the gardeners and farm-labourers.

But incomparably beyond all other sources of delight, open to the uninitiated and hermetically sealed to the "Fancy," was the contemplation of a certain domestic scene which went on in the particular corner of the Hall assigned to the Pomeranian family. Father, mother, and a whole litter of Pups were here secured together in a sort of pen, or fold, from which there was no escape. Never was a better example of a certain kind of unwilling head of a family, than was furnished by that Pomeranian father. He was chained to his home, and so was his excellent consort; but while she lay contentedly in the midst of her offspring and completely covered and overwhelmed by the little wretches who were sprawling all over her, the sire was found sitting at the very extremest limits of his chain, and with his head averted from the group, in a kind of desperate attempt to ignore the whole concern. It was perfectly useless for any of the scions of his house to attempt to attract his attention. To him they represented doctor's bills, school and banker's accounts, and disturbed rest, and nothing else in the world; and when he raised himself on his hindlegs, and placing his paws upon the edge of the fold, gazed upon the world outside and uttered one drawnout melancholy howl, it was the most perfect satire on undomesticated Paterfamilias that canine reproof could administer. And the poor mother, too, left at home with all the trouble and all the labour devolving upon her, and looking as pleased and contented all the while as the other looked disgusted! It was as good as a sermon. Better, perhaps, than some.

Great monster boar-hound, alone worth a moderate journey to get a sight of; gigantic neighbor of the above, with your deep pointed nose, and your sable fur; sweet faced muff from St. Bernard, whose small intellect is what might be expected of a race living on a mountain with only monks for company; small shadowy-faced Maltese terrier; supple foxhound; beloved pug; a. hound; detested greyhound of Italy; otter-hounds that look like gamekeepers — each and all I bid you farewell, and proceed yet a little further on my way through the suburbs of North London.

Curiously enough, within a mile of that great dog show at Islington there existed, and exists still another dog show of a very different kind, and forming as complete a contrast to the first as can well be imagined. As you enter the enclosure of this other dog show, which you approach by certain small thoroughfares of the Holloway district, you find yourself in a queer region which looks, at first, like a combination of playground and mews. The playground is enclosed on three sides by walls, and on the fourth by a screen of iron cage-work. As soon as you come within sight of this cage, some twenty or thirty dogs of every conceivable and inconceivable breed, rush towards the bars, and, flattening their poor snouts against the wires, ask in their own peculiar and forceable language whether you are their master come at last to claim them?

For this second dog show is nothing more nor less than the show of the Lost Dogs of the metropolis — the poor vagrant homeless curs that one sees looking out for a dinner in the gutter, or curled up in a doorway taking refuge from their troubles in sleep. To rescue these miserable animals from slow starvation, to provide an asylum where, if it is of the slightest use, they can be restored with food, and kept till a situation can be found for them; or where the utterly use-less and diseased cur can be in an instant put out of his misery with a dose of prussic acid — to effect these objects, and also to provide a means of restoring lost dogs to their owners, a society has actually been formed, and has worked for some year and a half with very tolerable success. Their premises are in Hollingsworth Street, St. James's Road, Holloway, and it is there that the public will find a permanent dog show, of a very different sort from that which "drew" so well at the Agricultural Hall, Islington.

At the Islington dog-show all was prosperity. Here, all is adversity. There, the exhibited animals were highly valued, and had all their lives been well fed, well housed, carefully watched. Here, for the most part, the poor things had been half-starved and houseless, while as to careful watching, there was plenty of that in one sense, the vigilant householder having watched most carefully his entrance gate to keep such intruders out. At Islington there were dogs estimated by their owners at hundreds of pounds. Here there are animals that are, only from a humane point of view, worth the drop of prussic acid which puts them out of their misery. Now we are accustomed to think that with human beings, high feeding, luxurious living, and constant appreciation on the one hand, and want, privation, and contempt on the other, will produce certain results on the character. Will it be considered too great a stretch of the imagination to say that something of the same sort is observable in lower animals? As I sit and write I get a glimpse through my window of a certain populous thoroughfare. I see the cab-horse trot my way by, with his head down and his ears slightly back, in a sort of perpetual protest; and presently I see a couple of highly-groomed ponies dance past with curved necks, and ears pricked forward, and hardly touching the ground, which they seem to despise. Is it fancy to suppose that this is not entirely a physical matter, and that there is something of arrogance about these spoilt beauties, and of humility in the poor cab-horse? Was it purely an over-indulged fancy that made me discern a great moral difference between the dogs at the Islington Show and those at the Refuge in Holloway?

I must confess that it did appear to me that there was in those more prosperous dogs at the "show" a slight occasional tendency to "give themselves airs." They seemed to regard themselves as public characters who really could not be bored by introductions to private individuals.

When these last addressed them, by name too and in that most conciliatory falsetto which should find its way to a well-conditioned dog's inmost heart, it was too often the case that such advances were received with total indifference and even in some cases, I regret to say, with a snap. As to any feeling for, or interest in each other, the prosperous dogs were utterly devoid of both. Among the unappreciated and lost dogs of Holloway, on the other hand, there seemed a sort of fellowship of misery, whilst the urbane and sociable qualities were perfectly irresistible. They were not conspicuous in the matter of breed, it must be owned. A tolerable Newfoundland dog, a deer-hound of some pretensions, a setter, and one or two decent terriers were among the company; but for the most, architecture of these canine vagrants was decidedly of the composite order. That particular member of the dog tribe, with whom the reader is so well acquainted, and who represents an important family of the mongrels, in all his-absence-of-glory. Poor beast, with his long tail left, not to please Sir Edwin Landseer, but because nobody thought it worth while to cut it, with his notched pendent ears, his heavy paws, his ignoble countenance, and servile smile of conciliation, snuffing hither and thither, running to and fro, undecided, uncared for, not wanted, timid, supplicatory — there he was, the embodiment of everything that is same pitiful, the same poor pattering wretch who follows you through the deserted streets at night whose eyes haunt you as you lie in your bed after you have locked him out of your house.

To befriend this poor unhappy animal a certain band of humanely-disposed persons has established this Holloway asylum, and a system been got to work which has actually since October, 1860, rescued at least a thousand homeless dogs from starvation. The modus operandi adopted and recommended by the committee of this remarkable institution for preventing the poorer London dogs from going to the dogs, is simply this: If it should in the course of your walks about the metropolis that that miserable cur which has been described above should look into your face and find in it a certain weakness called pity, and so should attach himself to your boot-heels; if this should befall you, and if you should prove too feeble a character to answer the poor cur with a kick, you must straightway look for some vagrant man or boy — alas! that are common in this town as wandering dogs — and propose to him that for a certain guerdon he shall convey the dog to the asylum at Holloway where he will be certainly taken in, and receipt handed to the person who delivers him at the gates. It is not, upon the whole, considered a good plan to remunerate the man to whom the vagrant dog has been given until his part of the Contract has been performed and this same receipt has been obtained. For, in the archives of the benevolent society whose system we are examining, there are recorded cases in which credulous persons have handed over the dog and the reward together to some "vagrom man," (sic) and somehow the animal has never found its way to Holloway after all.

Once at the "Home," the dog has a number tied round his neck similar to those which are appended to our umbrellas at the National Gallery, and which number corresponds with an entry made by the keeper of the place in his book, stating the date of the dog's arrival, and describing his Breed—if he has any—and, at all events, his personal appearance as far as it is describable.

The dog's individual case is then considered. If he be ill and his life be obviously not worth preserving he is humanely disposed of with a little prussic acid. If on the other hand, there seem some reasonable prospect of his obtaining a home hereafter, or if he appear to be of some slight value, he is doctored, fed, and gradually restored to health. Dogs are sometimes brought to the asylum in a most piteous state of exhaustion, and sometimes one of these poor little things will, after receiving a carefully administered meal, curl himself upon the straw and go to sleep for twenty-four hours at a stretch. From the greatest depths of prostration they are recovered by judicious treatment in a wonderfully short space of time. The society has also employed persons occasionally, to go about the streets and, in extreme cases, to administer a dose of prussic acid to such diseased and starving dogs as it has seemed merciful to put a quick end to.

Now, really, among all the queer things which a man might devote a whole lifetime to routing out and which lie within the limits of this metropolis, the existence of such an association as this is one of the queerest. It is the kind of institution which a very sensitive person who had suffered acutely from witnessing the misery of a starving animal would wish for, without imagining for a moment that it could ever seriously exist. It does seriously exist, though. An institution in this practical country founded on a sentiment. The dogs are, for the most part, of little or no worth. I don't think the Duke of Beaufort would have much to say to the beagle I saw sniffing about in the enclosure, and I imagine that the stout man, who owned the smaller terriers at the show, would have had little to say to the black-and-tan specimens, which mustered strong in numbers, but weak in claims to admiration, in the shut-up house in which there were as many lost dogs as in the enclosure outside. The thing owes its existence, as has been said, to a sentiment. It asks for but a very small donation, and does not enter into competition with those charities which would benefit the human sufferer. The "Home" is a very small establishment, with nothing imposing about it — nothing that suggests expense or luxury. I think it is rather hard to laugh this humane effort to scorn. If people think it wrong to spend a very very little money on that poor cur whose face I frankly own often haunts my memory, after I have hardened myself successfully against him, if people really do consider it an injustice to the poor to give to this particular institution, let them leave it to its fate; but I think it is somewhat hard that they should turn the whole scheme into ridicule, or assail it with open ferocity as a dangerous competitor with other enterprises for public favour. I should be slow to believe that the five shillings which is sent to the Holloway asylum is taken away from the poor, or that for the want of it some deserving mechanic, with his wife and family, will actually "go to the dogs." At all events, and whether the sentiment be wholesome or morbid, it is worthy of record that such a place exists; an extraordinary monument of the remarkable affection with which English people regard the race of dogs; an evidence of that hidden fund of feeling which survives in some hearts even the rough ordeal of London life in the nineteenth century.

7

Moose

Oh, Moosie Prince, oh, Moosie Prince,
You are a very good doggy.
Your feet are big, your nose is long,
your brain is large, your heart is strong.
Oh, Moosie Prince, Oh, Moosie prince,
you are a very good doggy.

Mack was a bad sheepdog, and I might have sold him if I hadn't believed his new owner would lose patience and put him down. Mack was big for a Border Collie, sixty-five pounds, and his genes had knots tied in them. He was a natural-born hysteric; Mack wanted to work so bad, he'd rev himself up until smoke poured out of his ears, lay rubber straight to the sheep, and when they scattered he'd grab hold of one. Mack preferred to bite that portion of a sheep that was going away from him. When a ram is leaving in a hurry, the last part dangling behind is what makes him a ram, and that's what Mack bit. It's annoying to stitch up a ewe. Stitching a ram's testicles is arduous.

I didn't give up on Mack. I taught him his lefts and rights, taught him to stop — sort of. One day when I was bellowing at Mack, Ethel Conrad asked, "If you don't like that dog, why don't you give him to someone who does?"

My wife, Anne, said she'd take Mack. Right away, she changed his name to "Moose," which is short for "Moosie-Prince." She made up a song for him, sung to the tune of "O Tannenbaum": "Oh Moosie-Prince, oh Moosie-Prince, you are a very good doggy." When Anne did chores, she took Moose and he became the "Hold 'em" dog. In January, after an icy rain, when milking ewes are desperately hungry, the "hold 'em" dog keeps the sheep from trampling you into the muck as you pour grain into their feeders.

Moose was protective of Anne, and if any man had lifted his hand to Anne, the only person who'd gain from it would have been that man's beneficiaries.

Moose kept order amongst our dogs and was the official chore dog — the dog who stays home to do the farm work while the other dogs go off to sheepdog trials. Moose didn't have time to be goofy anymore.

One March, a blizzard closed I-81 for three days. Three days before the snow started falling, we'd sheared our ewes, and 180 naked sheep were wandering through the brushy rough ground downstream from our barn. I don't know why I didn't get worried until the evening weather forecast, when it was already snowing hard. It was getting dark when I finally decided to bring the sheep into shelter. I didn't expect any trouble — the sheep'd be bedded down behind the wind-break as usual, and Moose was nearest the door so I took him to help. I suppose the approaching bad weather had made the sheep restless, because they weren't where they should have been, weren't anywhere in sight, and solid blackness was coming down fast, like a theater curtain.

Our flock was out there somewhere in a hundred rough acres and there wasn't anything I could do about it. "Okay, Moose. Let's go home . . . Moose? . . . MOOOOSE." I whistled, I shouted, I used words the preacher warns us against. About the time I was ready to give up, over the hill they came streaming; all 180 of them. I don't know how Moose found them in the snowy night. I do know he brought them calmly and quietly, just like a real sheep dog.

Next morning, it was ten degrees and blowing snow over a two-foot base, and if the sheep hadn't been snug in our barn, we would have lost half of them to hypothermia and pneumonia. Thousands of dollars in sick and dying animals.

After that night, Moosie slept on a sheepskin pad beside our bed — on Anne's side. Every dog, even the bad dog, must have his day.

♠♠♠

Whatever cultural impulses produced early dog shows, their logic seems plausible. As Jno Speed wrote in *Harper's Weekly*,

Anyone familiar with the work which a Pointer is supposed to do in the field will see the wisdom of . . . [show points assigned to aspects of the Pointer] . . . perhaps except that of color which probably might have been left out of the account entirely and given over to the points apportioned for legs and feet. The Pointer must have great intelligence, great power of scent, must have great strength and speed so as to do his work thoroughly and rapidly. The Pointer therefore structurally perfect according to the standard is very likely to excel in those qualities most required.

A century later, the AKC clings to this logic. If a Lassie collie is expected to work all day it should, according to the breed's standard, "carry no useless timber, standing naturally straight and firm. The deep, moderately wide chest shows strength, the sloping shoulders and well bent hocks indicate speed and grace and the face shows high intelligence."

In 1866, British sportsman John Henry Walsh (a. k. a "Stonehenge") systematized (and/or created) these conformation standards, assigned point values to those aspects he thought most important, and at dog shows afterwards, in Britain and in the United States, actual living pointers and Lassie collies who most closely resembled the judge's mental interpretation of Walsh's notions were dubbed "Champions." For all practical purposes, Walsh designed many of our modern dogs. No individual has had such influence before or since. It is only fair to ask how much Stonehenge actually knew about the breeds he described. (See Appendix B.)

♠♠♠

In the 1870s, the members of the Westminster Breeding Association bred, imported and hunted English pointers and setters. (Their imported "Sensation" — known as "Don" — was a field trial and bench pointer who won $1200 his first year in the United States. That's Don on Westminster's logo today.)

The actual management and training of their dogs was done by servants, some experienced British trainers imported for the job, others servants who'd failed as housemen or coachmen. Come the day of the hunt, the dogs appeared, were shot over, and, as the owner returned home, wrapped in a lap robe, enjoying his flask or cigar, he might be forgiven for thinking that he understood quite a lot about dogs and could evaluate them perfectly well. Who was to gainsay him? His kennel man? His dog breaker?

In 1876, the Westminster members put on a dog show in Philadelphia, repeated it the next year, and, emboldened by two successes, incorporated as the Westminster Kennel Club. In 1878, they rented a venue, solicited entries, and put on what came to be called "The New York Dog Show." It was a smash hit, and the two day show was extended to four to accommodate curious spectators. Although Westminster was not the only American Kennel Club (New England, Maryland, and Cincinnati clubs were also prominent), Westminster was indisputably the most important Kennel Club in America.

Francis Butler described the New York Dog Show thus:

The great bench show was held in the Hippodrome which had been fitted up expressly for the purpose. Stalls were erected around the capacious arena for the accommodation of the dogs; but the entries were so much in excess of the calculations that extra stalls were built, at the last moment, inside the arena. There were also two rings into which the several classes of dogs were taken to be judged. The show opened Tuesday, May 8th. As early as ten o'clock Monday the dogs

began to arrive. They came by all sorts of conveyances. Some
were packed in huge coops marked "with care," others were
led by stout iron chains, and still others were carried in bas-
kets or in the arms of their owners. Among those bringing
their pets for exhibition were many elegantly dressed ladies.
Crowds of men and boys surrounded the entrance. Some of
the dogs were disposed to be quarrelsome, especially the big
fellows, and many times the crowd scattered with ludicrous
haste at the unexpected growl of some ferocious-looking
brute. There was danger from some of them, too, for their
owners took great care to keep them at a safe distance from
the legs of imprudent bystanders. Not a few had great difficulty in holding the powerful animals in. Other dogs were
quiet and friendly, but not less annoying to their masters by
plunging about and entangling their chains in seemingly
inextricable confusion.

The spectacle inside the arena, when everything was in
readiness, was very attractive. There were over eleven hundred entries of all classes, from the huge Siberian blood-
hounds, the magnificent St Bernard dog, the Newfoundland
and the mastiff, down to the most delicate toy dogs. To the
latter were devoted several stands in the centre of the arena,
and this was one of the most attractive spots in the show. The
little things were rigged up with ribbons, wats, cushions,
bells and lace collars, in the nicest dainty style. Two large
pups with lace collars were very amusing. Others were the
occupants of a number of mahogany-framed glass cases. One
of these, a tiny mite of a thing, with long silken hair, bore the
ferocious name of "Danger." There were also three beautiful
Italian greyhound puppies, five delicate Japanese puppies,

and six little white balls nestling under their Blenheim spaniel mother. The principal attendance was during the evenings, when the building was crowded to its full capacity. There were quite as many ladies as gentlemen present, and they seemed to take quite as much interest in the dogs. The only drawback to the enjoyment of the show was the dreadful howling that filled the building and at times almost prevented conversation. Mr. Bergh's speech on Tuesday evening was inaudible six feet from where he stood. The larger dogs were, as a rule, dignified and quiet; but the petted darlings of the drawing-room expressed their anguish over their imprisonment and loss of home luxuries in tones that must have pierced the very hearts of their fair owners.

The show was in every sense a great success, and will probably prove to be the first of a long series of such exhibitions. It was held under the auspices of the Westminster Kennel Club, and for a first enterprise of the kind, the management was noticeably free from annoyances and mistakes. It lasted four days, and every one who visited it was delighted and entertained. But if the question of holding another bench show were left to the dogs, it would doubtless be rejected by a large majority.

The Westminster Kennel Club's Long Island clubhouse boasted a banquet-sized dining room, billiard and card rooms, and ten bedrooms. As many as 200 dogs and puppies were kenneled for members shooting pigeons on the spacious grounds. On June 21, 1884, the Westminster Kennel Club invited other kennel clubs to join them at Delmonico's restaurant "to discuss the propriety of uniting in a general association with the object of securing uniform rules for the conduct of bench shows, adoption of standards . . . and such other kindred matters."

At a Virginia garden party I met a Virginia horse vet who'd joined the Westminster Kennel Club when his college roommate invited him. "They don't know anything about dogs," he confided. "They're just in it for the parties."

♠♠♠

Although I'd seen Border Collies at a show in Brisbane, Australia, the 1993 Westminster Kennel Club show was my first American dog show. The Westminster show is at the core of America's dog fancy, its Grammies, Emmies, and Academy Awards. During Westminster, extravagant black-tie dinners honor dog world celebrities; dog food manufacturers host hospitality suites; and dog writers (the Dog Writers Association of America) hold their annual meeting/awards ceremony. The *Washington Post* assigned me to write about the show.

That February had seen heavy snows, and our train chugged towards New York City through brilliant white countryside while we passengers complimented ourselves on our foresight. Roads were horrible, the airlines struggling to transport baggage and passengers.

In New York, in Madison Square Garden's press room, I produced my credentials for a woman who was too busy reducing an underling to tears to pay me any attention. Usually press officers are courteous and helpful; this press officer was neither. When I asked her about the Westminster

Kennel Club, what did it do besides hold this show, she opened the show catalog to the WKC officers and tapped it.

"But what do they do?"

She tapped the names again.

I said, "Okay. What do I do to join."

"Oh, you couldn't do that!"

The Dog Writers Association of America (DWAA) awards dinner was held that evening at Keene's Chophouse, and the keynote speaker was the same publicist, Ms. Thelma Boalbay. If she had given a moment's thought to her speech before speaking, it wasn't apparent. She rambled on about other Westminsters when the weather had been bad. She dated these Westminsters by saying "that was the year the boxer won, you know the one shown by Jane Forsyth. You have to get up pretty early in the morning to beat Jane Forsyth" or "that was the second year that springer took it. Clint Harris is such a clever handler." Ms. Boalbay went on to make it clear that not any-old-body could obtain a WKC Press blue ribbon (which permitted nominally better access, but conveyed higher status); consequently, most DWAA members must be content with the humbler red. She thanked the dog writers for ignoring embarrassing flaws and glitches at previous Westminsters, and the dog writers applauded their discretion.

I'd been a professional writer for fifteen years but had never read the writers I met on that occasion, though some were, I was told, famous and influential. I was introduced to publishers from publishing houses I didn't know existed. The DWAA's top awards are named after a famous writer called

"Maxwell" — whoever he was. The dog writers were pleasant folks. Keene's food was okay.

The next morning I was guided through the dog show by a woman who had admired *Eminent Dogs*. She told me which dog would win tomorrow and that the winner had been decided months ago (she was wrong). She said that cosmetic surgery was commonly used to improve show dogs and inspected one Cairn terrier's ears skeptically.

In Colorado (she said), a show malamute pup had everything it took to become a champion except two testicles — one had not descended. Since undescended testicles are a disqualifying fault, the breeder located a vet who stretched a little skin, inserted Styrofoam, and replicated the item nature hadn't provided.

Some six months later, sans fanfare, the real testicle descended and at a show, the judge was feeling around under there, counting: one, two . . . three?

Perhaps my guide's claims, her "insider" knowledge, were a natural consequence of the arrogance and secrecy of the dog fancy. I couldn't help recalling the man at Frying Pan Park who'd been so sure Border Collies were "already in."

I wondered who was paying to rent the Garden. It wasn't spectators — the bleachers were mostly empty. Corporate sponsors? The AKC?

♠♠♠

At any AKC dog show, the dogs come out on lead and are paraded before the judge. Next they stand for the judge's inspection. Finally, the handler trots around the ring so the judge can evaluate the dog's gait. The judge considers, sometimes reinspects, dismisses those dogs that haven't placed, and arranges the winners. Ribbons and trophies are awarded. Dogs are evaluated by breed (all collies), next all the breed winners are judged by group (all herding dogs), and finally the winners of these eliminations are judged against each other for the top honor, "Best-in-Show."

Top show handlers are brilliant with dogs. In *Born to Win*, Patricia Craige writes that "there is something almost magical in the bond between a great handler and his great dog. It is a true bond. They are in tune and connected. Neither of them makes a mistake as they work together to cover whatever shortcomings do exist and feature all the virtues." Fortunately, those owners less gifted with dogs or too rich or too busy could hire the dog work done.

On the floor of Madison Square Garden, three show rings were going at once, and claques cheered their favorites and the dogs barked. In back, where the dogs were held until their turn in the spotlight, it was a madhouse. Against one wall concessionaires sold dog gear and dog periodicals. There were long rows of dogs—hundreds of crated dogs stacked one on top of another while spectators squeezed past down the aisles.

All those strangers, nowhere to run. I wondered how the dogs put up with it.

Most of the people waiting to show their dogs were middle- and upper-middle class, most did dog shows as a means of connecting to their pets. The more important breeders tended to be wealthy, and the wealthiest wouldn't show up until five minutes before their dogs were taken into the ring. Some owners only visited their dogs at the biggest shows.

At the 1994 Kal Kan "Top Show Dogs of the Year" dinner, winning owners included Mrs. Firestone, Avis Rent-A-Car's CEO, and a gentleman representing a Japanese consortium that had campaigned a dog long distance. The black-tie optional event had open bars, excellent entrees and wines, and a dance for two hundred guests. Since Hershey Foods had just bought Kal Kan, dog fanciers fretted that Hershey might do a cost-benefit analysis of this. The only obedience dog among the winners was an (ILP) Border Collie.

In a demarcated zone, professional handlers and groomers congregated with clippers, hot combs, and touch-up dyes — a regular assembly line of dog preparation. Since many handlers believe that training diminishes a show dog's spontaneity, one had to watch where one stepped.

I asked a sheltie owner why he bred show dogs. "It's an art form," he said. "It's like painting with genetics."

Not unlike dog fanciers, sheepdog trainers imagine they can know much about a dog simply by looking at it. A puppy with his tail held high and frivolous, one who doesn't attend to its owner and even disregards him; when that dog's eyes can't hold a thought for a moment, there's trouble ahead. But every sheepdog trainer's judgment is tentative, for when the dog goes to work with sheep it may (and often does) lose its ill manners and puppy behaviors to become a new kind of creature: a "sheepdog."

The dog fancier believes that one can judge — absolutely — an Australian shepherd's abilities while the dog is on leash in a forty foot show ring — its handler enlivening it by disgorging chewed liver from his/her mouth to the dog (an unsanitary practice known as "baiting").

Eighteen Australian shepherds debuted that year at Westminster, and of the dogs shown, they were most like Border Collies. The show Aussies were jumping all over their owners and those spectators unlucky enough to be at ringside. They were fat but cute. I imagined I saw a gleam of sense in one calmer dog's eyes, but the judge didn't agree with me.

I asked one winner if her dog worked livestock. "Sure he could," she said. "It's what he was bred for. But I don't do it. He might get hurt."

♠♠♠

I don't believe the American Kennel Club comprehended that the Border Collie community didn't wish to have anything to do with them.

When Dr. Kissinger boasted "power is an aphrodisiac," he named only one of power's intimacies. How many Americans would refuse a White House dinner invitation or a chance to socialize, however briefly, with a television or movie star?

After a hundred years of solemnly weighing dog clubs desperate for AKC recognition, AKC officials were unable to conceive that a perfectly good club like the USBCC might believe the AKC was at best, silly, and at worst, harmful to dogs.

After Jack Ward's wife was diagnosed with cancer, he resigned as chairman and James Smith, a New York State industrialist, was elected to replace him. Shortly thereafter, Ethel and I got a big purple book in the mail, a guidebook for aspirant breed clubs seeking AKC recognition. It was expensively produced for a publication so infrequently used (power is, or must seem, indifferent to money) and was authoritarian in tone. The big purple book assumed that we, who had argued so strongly against breeding our dogs for dog shows, suddenly wished to whisk them into the ring, that we were willing to allow the AKC to make big decisions about our dogs, and indeed, our lives, in so much as our lives were intertwined with our dogs. It was not an appealing invitation.

The cover letter accompanying the big purple book was blunter. Over the name of a junior AKC official it said we had thirty days to start aspiring, or they'd find somebody who would.

♠♠♠

After the 1878 New York Dog Show, the dog fancy flowered, and dog shows proliferated. In cramped show conditions, distemper and other communicable diseases were rife. After forty dogs died at or immediately after the 1883 New York show, one correspondent wrote

> My experience at shows, whether held at New York or any other place, is that they are injurious to the dog and in nearly every case sickness, sometimes of a mild nature, sometimes of a serious nature, but almost always sickness of some kind. No immature dog should be exhibited, puppy classes should be

done away with. I have never visited a show without detecting sickness, sometimes distemper, often severe colds, etc. It is my opinion that it is impossible to assemble a thousand dogs under one roof and not have a few sick dogs among the number, and when the sickness is of a contagious nature, it can easily be understood why dogs return sick from the shows.

Dog show judging was capricious, favoritism commonplace, even outright fraud. In 1902, Charles Henry Lane wrote that

it was formerly not uncommon in some breeds, where the dogs were much alike, for a good specimen to be shown, say on Tuesday, taken out that night, of course a substitute brought in on the Wednesday morning, shown again somewhere on the Wednesday, and on some occasions, where dates and distances permitted, at two or more shows during the week, either no name or pedigree being given, or a different name at each show, and, the same owner having a number of dogs of the same colour, "the changes were rung" to suit the circumstances of the case!

In those days too, it was not safe to claim a dog at a show unless you had some positive means of identification or you were more than likely to find a very inferior animal in his place at the end of the show!

Lending and borrowing dogs were everyday occurrences, and if an exhibitor found a dog entered (as the questions of age, color, name and even sex, were treated in the most free-and-easy manner) was not very 'fit' or in good form when the show came off, he would substitute another of his own (or somebody else's) instead.

In 1876, The National American Kennel Club was formed "to promote, encourage and improve the breeding of a superior class of dog . . . to publish a studbook for the registration of pedigrees and to adopt rules and regulations for conducting Field Trials and Bench Shows."

The NAKC was slow to publish its promised studbook, so *The American Kennel Register* began registering dogs. Complaints were numerous, viz: "Please make some inquiries about pedigree of Bo. 1103. It reads: dam - Beauty by imported Racer,' etc. If this is Robert Walker's Racer, it is wrong, and I am prepared to prove that nothing is known about Racer's pedigree. The pedigrees of no. 1104 and 1107 are wrong in re Topsy. It should be the same as the dam of Fannie (383). I know all about Topsy."

Topsy wasn't the only dubious dam. With alarming frequency, grinning their impure grins, mongrels slunk into these early registries.

Duplication of names was another problem. How could there be two registered "Queen Bess's," one in Cleveland, the other in Bethel, Maine?

Because the NAKC was dominated by bird dog field trial enthusiasts from the South and West, Eastern bench show people were dissatisfied.

That first organizational meeting called by the Westminster Kennel Club fizzled, but the clamor for an overarching ruling body was overwhelming. There were disputes whether this ruling body would be comprised of delegates from different clubs or individual members—Westminster holding for the former—but there was no doubt that American dog fanciers wanted a strong central government to tell them how to behave.

♠♠♠

The arrival of the big purple book came almost as a relief. The other shoe had dropped, the long-anticipated attack had come: war had been declared.

Although there were other important venues for those who wished to "show" their dog in obedience, AKC sanctioned events were most numerous, and many obedience competitors coveted AKC titles which successful competition appended to the dog's pedigree. That this string of initials is similar to the string of letters once appended to Lord Mountbattan's name is perhaps not coincidental.

Border Collie obedience handlers faced (AKC reps whispered) an unhappy choice. Either (a) help get the Border Collie fully recognized, or (b) the AKC would kick the Border Collie out of the Miscellaneous Class so that the dogs would no longer be able to compete in obedience. To people whose lives revolved around these competitions (and Border Collies are wonderful obedience dogs) it was an excruciating dilemma. Obedience people weren't interested in dog shows, and most knew that full AKC recognition would damage their breed, but not being able to compete any more was unthinkable. (In Canada, only one registry can register a breed, and the matter is decided by owners' vote. When the

Border Collie community overwhelmingly preferred the Canadian version of the ABCA to the Canadian Kennel Club, that organization retaliated by booting Border Collies out of its obedience trials. As I have noted, any dog, any breed, any papers or none can run in a sheepdog trial.).

Although they and their sport were the ugly stepchildren at dog shows, obedience people and their skills weren't overvalued by sheepdog people either. A few obedience handlers had moved into sheepdog trials, but these excepted, I don't know any sheepdog trialist who has ever bothered to attend an obedience competition.

Obedience competitors weren't numerous — probably fewer than 200 of a Border Collie community of 10,000 — but they were educated, had one foot in the AKC camp already, and most were scared enough to cooperate. Some believed recognition was inevitable, and if they were the AKC breed club, they could emphasize the Border Collie's working abilities and minimize the damage the AKC would do to the breed.

It's easy to seize the high moral ground here: what kind of person would put his or her hobby ahead of a dog breed's welfare? Alas, if the AKC controlled dog trialing as they control obedience trials, I can't think we dog trialists would have been as brave as we were.

At a distance, the AKC's power seems chimerical. Some dog fanciers are wealthy, some are prominent in the land. Among their number they count corporate CEO's, investment bankers, attorneys who represent Fortune 500 corporations. How can a mere dog show organization hold power over people like these?

AKC authority doesn't derive from affection. Through the long years of the dog wars, I never heard one dog person, obedience or conformation, confess fondness for the AKC. The highest praise I ever heard was said—in response to criticism by the Humane Society of the United States—"The AKC isn't much, but it's ours."

Whenever I came to an AKC event, dog fanciers took me aside, to relate scandals.

But they wouldn't be seen with me publicly. When I suggested they might change the AKC, make it more responsive, more useful to dogs, some said nobody can fight city hall, most were startled—as if I'd suggested dismantling Mount Everest and reconstructing it bottom to top in Neptune, New Jersey.

To these dog fanciers, the AKC was a force of nature: capricious, arrogant, sometimes manipulatable by flattery. Its monied momentum swept reformers away in the flood.

AKC dissidents face real sanctions. Their dogs stop winning, coveted judging assignments go to others, they are shunned by their fellow fanciers. Dog writers don't get assignments, don't get published, can't get interviews, and never, ever get a Blue Ribbon Press Pass at Westminster. The ultimate sanction is the sanction Border Collie obedience people faced. Should the AKC decide you are an enemy, you will no longer be able to enter their events and earn those coveted titles. Overnight, you find yourself the owner of some very expensive mongrels. Dog fanciers sometimes whisper, "The AKC can put you out of dogs."

♠♠♠

We knew ours would be an uphill fight but hoped to generate enough bad press to force the AKC to negotiate. Ethel Conrad wrote Chairman Smith asking for a meeting, we reopened the Border Collie Defense Fund, and we started calling reporters.

As it happened, the AKC had bit off more than they could chew. Besides moving against Border Collies, their Directors had determined that Labrador Retrievers had become too short, that longer-legged Labradors looked more like "working dogs," so they commanded that short-legged labs couldn't win dog shows. Owners of short-legged Labs banded together to sue the AKC.

And while the AKC had been pursuing us, they'd also been after the Cavalier King Charles Spaniel, toy spaniels historically registered by the Cavalier King Charles Club. Cavs are so cute the registry feared massive breeding by puppy mills and had an ethical policy that if any member's pup was discovered for sale in a pet shop, the breeder either (a) had to produce a pretty good explanation or (b) they'd lose registration privileges. This sensible policy kept the Cavs out of the pet shops. When they got their own big purple book, the Cav club fended off their unwanted suitor by saying they needed more time to think things over. They dilly-dallied as long as they could, praying for a change in the climate at 51 Madison Avenue. When they finally put matters to a vote, 91 percent declined the AKC's advances. So, as they had with Border Collies and Australian shepherds, the AKC sought out a handful of Cavalier King Charles owners to become the official breed club.

Without wishing the short Lab or Cav owners ill, we were glad the AKC was overreaching.

Dogs are not news unless they have savaged a child and photographs are available. Dog stories are filler material reserved for slow news days and (by long custom) a careful ironic distance is preserved between the newspaper and dogs. Every headline must be a bad pun ("Fido is Unfaithful, Dog Gone It"; "A Woof in Time Saves Family") and so on and so on. The *Post*'s article about the AKC herding clinic had been no exception: "As Faction Pushes for AKC Approval, Owners Protest It Is Not Better to be Seen than Herd."

The Dog Writers Association of America was founded in the 1920s by reporters who covered dog shows for the society pages. Many DWAA members were in thrall to the AKC and never realized how routinely they self-censored their writing about that organization. Dog writers who told me such amusing stories of AKC dog ignorance and indifference never told them in print. The major dog magazines (who depend on dog show advertising) were circumspect. Freelance journalists who own dogs and write for general interest periodicals find dog articles hard to place. Many TV reporters and producers (and most of their bosses) live in greater Manhattan, and if they have animals they're likely to be cats.

Interesting the news media in our troubles wouldn't be easy, and three dog clubs quarreling with the AKC was a bigger story than one. Fortunately, Congress was on summer recess, so reporters had extra space to fill.

Charles Krauthammer owns a Border Collie. As an MD, he understands genetic arguments, and in his syndicated column, fired the first shot in July of 1994:

> Last month, the American Kennel Club, the politburo of American dog breeding, decided to turn the world's smartest dog, the Border Collie, into a moron. Actually it voted 11–to–1 to begin proceedings to turn it into a show dog which will amount to the same thing. A dog bred for 200 years exclusively for smarts will now be bred for looks. Its tail, its coat, its ears, its bite, its size will have to be just so. That its brains will turn to mush is of no consequence.

US News and World Report picked up the story and, with evidence that other journalists saw this story as real news, I approached *The New York Times* with an Op Ed piece:

> It has been a bad summer for the American Kennel Club, the organization that oversees registrations and dog shows for 130 U.S. breeds. It has been sued for $11 million by Labrador retriever breeders who resent its distaste for short retrievers; the two independent breed registries that the club sought to bring into its fold are fighting its efforts, and according to the *Washington Post* the Justice Department is reviewing allegations of antitrust activity. Conversely, it has been a great summer for the American dog.
>
> The American Kennel Club was formed in New York in 1878 by a group of rich dog fanciers who didn't let any old mutt into their club; breed registries aspiring to join had to turn over spotless stud books (the records of all purebred

matings within the breed) and produce an established con-formation standard (what the dog should look like). Those breeds that didn't measure up, or whose owners didn't care to become vassals of the kennel club, were ignored and considered mongrels. Yet as the club has grown in influence and its staff has swelled, it has become more concerned with Byzantine turf wars than with the welfare of dogs. Faced with declining registrations in recent years, the club raised its fees and cut its spending on education and health research by more than 60 percent. And it no longer requires spotless stud books as requisite for membership.

Although the club's prestige depends on the dog shows it oversees (especially the Westminster Kennel Club Show in New York City), most of its $29 million annual income comes from fees charged for registering puppies, most of which are household pets who get nothing from the club but a piece of paper with a number on it.

Most owners mistakenly think that these pedigrees are a warranty of quality. The American Kennel Club makes no such claims. It cannot guarantee that your new puppy is descended from the Champion Le Grande Doggo on its pedigree. Nor will it claim that registered dogs are any more sound than a mongrel you might pick up at the pound.

And it takes no responsibility for genetic problems in the breeds it oversees, holding that registries don't ruin dogs, breeders do. Thus the club's leaders wash their hands of the pet collie that goes blind and the beloved German shepherd that has to be put down at age 6 because of his hip dysplasia.

The kennel club has other problems. A group of Labrador retriever breeders are suing because the club decided this year that tall Labradors meet the breed standard but short Labradors don't. Owners of champion short Labs (worth up to $30,000) whose value suddenly plummeted were naturally displeased.

The club is also trying a hostile takeover of the King Charles spaniel, a fluffy, loving lap dog. The breed's independent registry, the Cavalier King Charles Club, has an ironclad rule that ousts any breeder who sells to a pet shop or a wholesale "puppy mill." When the American Kennel Club invited the Cavalier Club to join its ranks, the spaniel owners balked because the larger group accepts registrations from puppy mills. When 91 percent of the Cavalier Club voted against joining, the American Kennel Club encouraged the dissenting 9 percent to form a new breed group under the kennel club's control.

Or consider the Border Collie. Unrivaled as a working stock dog for farms and ranchers, it is a terrible pet because it is such a workaholic. Bred for its herding ability, it comes in all sizes, colors and shapes, and is thus not suitable to the show ring. Yet the kennel club (against the wishes of the Border Collie community) is trying to bring the breed into its "herding group" category. It is heedless of warnings from many prominent geneticists that breeding the Border Collie for shows will inevitably ruin the breed's working ability.

The club's single-minded show dog orientation, combined with its unwillingness to take responsibility for the health of the breeds it controls, has created witless Irish setters, sickly Akitas, cocker spaniels that can't hunt and German shepherds that can't walk. "I get two or three phone calls a day," dog trainer Vicki Hearne told me recently. "The people tell me about all the show champions in their dog's pedigree but that it keeps on biting the kids."

Breeding away from defects is complex, but it is not rocket science. To enter the Westminster Kennel Club Show, a dog must be a kennel club champion. Eye defects are estimated to affect 90 percent of rough collies in the U.S. Think what it would mean to the future of that breed if Westminster required an eye test.

Throughout the dog community, the Labrador retriever lawsuit was greeted with a muted cheer. If the AKC will not change to better represent the dog community and our wonderfully divergent American dogs, we would be better off without it. (*New York Times*, August 3, 1994)

After the *New York Times* weighed in, the story was picked up by other newspapers, and a *Time Magazine* reporter came to 51 Madison Avenue to interview AKC officials.

Though Border Collies were vital to us, surely they didn't matter much to the AKC. Registrations wouldn't be numerous, and though it may have been slightly embarrassing that the best obedience and herding dog wasn't "theirs," neither were the best sled dogs or bird dogs.

In the wake of scandals which had nearly wrecked the United Way, some conservative congressmen were reexamining charitable tax exemptions; the AKC grossed $29 million from its near-monopoly on purebred dog registrations, spent $2,000 every day just to maintain its employee pension fund, and paid no income tax. We believed the AKC was vulnerable to antitrust action, that the AKC's income tax exemption was shaky, and that rather than endure too much public scrutiny, they'd make a deal with us.

If they'd continue the present system where the Border Collie could compete in obedience but not the show ring, we'd support an increase in their fees for Border Collies. Reluctantly, we were even willing to let them start registering our dogs, so long as they didn't impose a conformation standard and start showing our dog in dog shows.

Under these circumstances the USBCC might even be willing to be the AKC parent club. (I can offer no better evidence of the USBCC's commitment to the Border Collie than its willingness to entertain this dismal compromise.).

Our hole card was silence. If they left our dog alone, we'd stop telling the world how wicked they were. Nope, not a noble position. Free speech is noble; I can touch my dog.

♠♠♠

One AKC Director, Jay Phinizy, voted against the takeovers on the simple grounds that "If they don't want to be a member of our club, why force them in?" Phinizy, a Scottish deerhound breeder, was known as a reformer. USBCC Director Sally Lacy invited Phinizy to her farm to see the dogs work. He advised her to quit trying to argue what was best for the Border Collie and persuade the AKC that recognizing the dog would be bad for the AKC.

A Richmond Op Ed editor phoned me to say they were reprinting my *New York Times* piece. He said the *New York*

Times had received a tremendous volume of mail about the piece, they'd been astonished so many people cared.

And, incidentally, the editor wanted me to recommend a dog. His family had always had mutts from the pound, but now, with the kids off to college and a little more time (and money) they thought they'd get something better: a purebred. What AKC dog would I recommend?

♠♠♠

The summer was too wet to make hay, and we were 700 bales behind schedule. I took Moose out for a training session. At ten years old, he was still hopeless, but I trained him anyway; he yearned for it. I couldn't slow him down and when he came on fast he got into serious trouble, but when we were done he danced on his back feet and wanted to nip my legs: he was beside himself with joy.

The telephone seemed connected directly to my central nervous system and jangled all the time. (Our phone no longer "rang," it "went off.")

I was told the AKC recalled its field reps to the New York office for a briefing.

We'd written each of the 500-plus AKC delegates but had not received a single response. Nor, excepting Phinizy, had any AKC director asked to see our dogs.

Ethel got a note saying AKC Board Chairman James Smith would meet with us at 51 Madison Avenue on August 10th.

We'd learned a little since our first AKC encounters. We assumed they'd be angry and would try to bully us again. Although we knew better, we still hoped the welfare of a unique dog breed mattered to them — at least that they'd pay lip service to that novel notion.

Cynically, we readied a two pronged argument, that recognition would harm the Border Collie *and* the AKC. Since Border Collies hadn't been bred to be morphologically "typey" (physically consistent), pups vary tremendously; even pups from the same dam and sire. "Predictability," the AKC likes to say, "is the best reason to buy a purebred puppy." Since show breeders breed for consistent morphological type, litters with long haired and short haired, flop eared and prick eared, red and black and white pups are a nightmare.

Furthermore, since there is no conformation standard and no dog show judge has ever seen or judged the breed, which dogs would he honor, which exclude? Is the conformation standard and judge's decision to be purely (and patently) arbitrary?

We found ourselves arguing that our dog wasn't good enough for dog shows.

Sally Lacy put together a video of Border Collie champions, demonstrating their morphological variety, and our handout contained a similar photographic montage. The handout contained a letter from geneticist Jasper Rine, who stated that show breeding and working abilities were incompatible, a letter from ISDS Secretary Philip Hendry describing what occurred in the UK when the Border Collie was recognized by the Kennel Club, a protest from Kay Guetzloff whose obedience Border Collie Sweep was the highest ranking AKC obedience dog in history, and a warning from Michael Fox of the Humane Society of the United States that

AKC recognition and subsequent breed popularity would inevitably put many dogs in animal shelters.

Lawyer Ray Mundy agreed to represent us again.

Same conference room. Same long table. Our team was Ethel, Ray, myself, and Sally Lacy, who had been involved with dogs all her life.

Their team was Chairman Smith at the head of the table, their new President Judith Daniels at the foot, and numerous vice-presidents in the middle. Jay Phinizy was relaxed, pushed back from the table—not quite part of either team.

After one terse nod, Smith attacked. He said he'd attended college in my part of the country, at Washington and Lee, where he learned an honor code which forbade stealing and abhorred lying. He went on about this at some length. Forcefully, without ever directly saying so, he implied I was a liar. Smith was a bony, angular man, used to getting his own way. He said the AKC had not increased its staff and that though their annual report showed a reduction in research funding, they hadn't actually reduced it.

I find bullying more hateful than confrontations. Fights fought years ago by a skinny kid in a tough Montana mining town made me who I am.

But before we went inside, we'd agreed that when the bullying started, we'd ride it out. Sometimes sitting still can be excruciating.

Brightly, Sally Lacy interrupted Smith's tirade, "Ah yes," she said, "My school had a motto. Harvard's motto is 'Veritas,' and we should all stick to 'veritas' here."

With great courtesy, Sally had topped Smith's school and stilled his rant. God bless Sally Lacy. When I poured myself a glass of water, my hands were shaking.

I said we came here to talk about the things we did agree about, that I was sure we all had the best interests of the Border Collie at heart. (Smith was right. I am a liar.)

Jay Phinizy disarmed tension with a couple quiet questions about the dog.

Ethel passed our handout to all the AKC officials. None opened it.

As it happened, Sally had known the AKC official who, way back when, drafted the AKC's interim Border Collie standard — without ever actually seeing a Border Collie. The staffer phoned Arthur Allen to ask what the dog looked like. Sally related this amusing history without noticing the gritted teeth.

I suggested the AKC might increase its fees for allowing Border Collies into obedience competition — without actually registering our dogs.

"Do you think we're only after money?" a staffer asked.

Another staffer asked how many members the USBCC had. "About 400," Ethel replied.

The staffer said that the "other club"(The Border Collie Society of America) claimed that many members.

I doubted their numbers, but it was easy enough to find out who was right: why not poll the Border Collie community? If a simple majority wanted to join the AKC, we'd no longer oppose recognition. We'd even pay half the cost of the vote, if the AKC were willing to be bound by the results.

Response to that suggestion? Zero.

President Judith Daniels asked a couple questions, and when Smith slipped into attack mode again, she suggested that the meeting would be more productive if we concentrated on those things we agreed on.

Sally Lacy said that since none of them had ever seen a working Border Collie they might not realize how morphologically dissimilar they were. She showed her video tape, arguing that the Border Collie wasn't good enough for the AKC.

Smith brushed it all aside. "You just don't like dog shows."

"I don't think there's anything wrong with dog shows," I said. "We've got a great dog show at our county fair. 'Best child's pet' is a real crowd pleaser because the dog can be any kind of old mutt and every kid gets a blue ribbon."

Ignoring their grimaces, I forged ahead. "I even judged a dog show myself once. It was at Sally's alma mater." (The Harvard Coop did it for a promotional stunt during one of my book tours.) "I gave first place to the shelter dog that had been oldest when adopted because that dog and its owner both deserved a prize. Second prize, I gave to the dog with the best name: Buck, a foxhound. Third prize, I gave to the young dog who should have been nervous being there, but he wasn't. Cool as a moose. No, I don't have anything against dog shows. Breeding for them is what bothers me."

Long silence followed my helpful disquisition.

Jay Phinizy asked if we would be willing to accept some sort of a "performance standard" for Border Collies in lieu of a "conformation standard." We said we would.

Smith asked again if we would accept a conformation standard and Ethel said no, that was not negotiable.

Judith Daniels said that they hadn't made any final decision, that all the facts weren't in.

"We'll provide you with any information you need," I said.

"We can negotiate everything else, but I believe you'll have to accept a conformation standard," Smith said.

Again, Ethel demurred.

One staffer wondered how many Border Collie registries there were and wrote down their names. (No AKC official ever contacted the Border Collie registries.)

I gave Smith and Phinizy copies of *Eminent Dogs, Dangerous Men*, explaining that it was a pretty good introduction to the Border Collie. And that was that.

I ran to catch the train back to Virginia, which was a weary train indeed. I tried to read. Couldn't. Tried to answer some of the correspondence I'd brought with me. Couldn't do that either. Tried to sleep. Couldn't. Was there something we could have done better? Should we have fought Smith instead of letting his rage roll over us? What on earth could we do now?

When I got off the train in Charlottesville at 10:00 P.M., I had a two-hour drive ahead of me. Climbing foggy Afton Mountain, steering between the little yellow lights in the blacktop, I kept asking myself the same witless questions over and over again, like my mind was running a squirrel cage. Twice I hit the brakes hard, when sixteen-wheeler hallucinations hurtled out of the fog.

♠♠♠

The short Labrador Retriever breeders had their day in a Virginia court lawyers call the "Rocket Docket" because they move cases so quickly. Venue was changed to New York City, where, months later, the short Lab breeders failed to get a preliminary injunction because, the New York City judge explained, the AKC are the dog experts. Thus far they'd spent—Labrador breeder Price Jessup told me—$200,000 dollars in legal fees.

I told Price our great advantage was the AKC's stupidity—contradictions and evasions in AKC culture had created a stupid organization.

"You've got it backwards," Price said. "It's actually a disadvantage. I've never lost a dime because of smart people. If the man I'm doing a deal with is smarter than I am, I'll hire somebody just as smart to negotiate with him. Stupid people have cost me tens of thousands. They act capriciously. They act against their own best interests."

♠♠♠

In the July-August 1994 issue of his *American Border Collie* magazine, publisher Jon Apogee wrote a hot editorial suggesting that we Border Collie people try to remove the AKC's income tax exemption. Jon sent copies of the magazine to every AKC Director, and Chairman James Smith flew into a rage. Apparently he tried to call Apogee, who wasn't home, then me, ditto; and when he reached Miss Ethel he shrieked at her, "Who is Jon Apogee! What do you know about Jon

Apogee!?" Ethel hadn't seen the magazine and hadn't a clue what Smith was yelling about. His psychic violence so shook Miss Ethel she could hardly speak when I talked to her hours afterwards. Though Miss Ethel was seventy, that was the first time she seemed frail.

By criticizing the American Kennel Club, we became the wild-eyed radicals of the dog world. A surprised Jasper Rine discovered mere genetic science daren't doubt the dog fancy after his students told him he was being denounced all over the Internet. When Jay Phinizy introduced Sally Lacy for a talk before Boston's Ladies Dog Club, he presented her as "one of those green-fanged Border Collie people."

But if we were seen as radicals by many, others applauded us — working dog people whose values had been derided by the dog fancy, veterinarians who had to treat the expensive, painful consequences of dog show breeding. Even some show judges congratulated us for standing up to the AKC. Jack Russell and Anatolian breeders phoned to ask how they could fight AKC takeovers of their breeds.

I believe that had a vote been taken of all the ordinary dog people who belong to AKC clubs, most would have sided with us.

The AKC found itself with more at stake than they had anticipated. I was told one AKC Director who voted against us confessed he agreed with all our arguments but didn't dare back down — AKC critics would be encouraged and Lord knows where that might lead.

♠♠♠

In September, Miss Ethel and I drove to the National Sheepdog Finals, held that year at the Horse Park in Lexington, Kentucky. Although Gael and Harry had qualified, I wasn't optimistic. Too many mornings, I'd been on the telephone when I should have been training. I went trialing as a relief from thinking about the AKC.

The first morning I sought coffee and donuts, skipping the handlers meeting where the judge gave instructions.

About ten am, Kent Kuykendahl had his Bill out examining the course.

"Good luck, Kent," I said. "Bill run soon?"

"No," Kent said. "We run this time tomorrow. I just thought I'd show Bill how it would be, the light, the shadows and all." He grinned, "If we're lucky the sheep will be running the same way tomorrow." (Kent and Bill came in sixth.)

I was too worried to think straight, certain that tomorrow's AKC Directors meeting was going to vote us in. Sheepdog trial? What sheepdog trial?

Miss Ethel had brought a portable television set, and that night, in her motorhome (a.k.a "Mumsie's Folly"), we watched ABC's 20/20 shred the American Kennel Club. I felt a little sorry for them. Show videos of horrible puppy mills, interview people who bought sick pups, and then ask what the registry is doing about it. How could the AKC prevent puppy mills from breeding too much without, in the same breath, preventing some of its rich and respectable breeders from breeding too much? How is a remote collector of begats checks going to know which dog registrations are fraudulent and which not? Although I thought the AKC had been whipped unjustly, I wasn't sorry they'd been whipped.

Next morning, we anxiously awaited to learn the AKC Directors' decision. About eleven o'clock, Judith Daniels called to say the Directors had kept us on hold. They wanted more information about the Border Collie. We assured her of our willingness, nay, *eagerness*, to provide it.

♠♠♠

Harry's run was in the early afternoon, usually not a good slot. When the sun's up, sensible sheep find shade and become contemplative. They get cranky when asked to change their schedule just because some sheepdog wants to parade them around a trial course. And Harry wasn't reassuring me: today, he was very much the lover; sniffing his sniffs, pestering Lewis Pulfer's bitch until she bared her fangs. There wasn't much I could do about it. I didn't want to ask Harry to run the National Finals five minutes after dashing his hopes for a love life.

The trial course was laid along a steeplechase course with a brush jump near the top and a water obstacle directly on the right-hand outrun. The dog couldn't see the water obstacle until he was on top of it and Harry didn't; his tail went straight up in the air like a drag chute deploying. He did stop, came left to get by the obstacle, and I gave him a small whistle for encouragement. The hill contour kept him from seeing his sheep, so Harry was running on hope and his memory of where they'd been when he set off. When Harry topped the rise and disappeared, I shaded my eyes waiting for him to reappear. There! There he is, coming on and I don't whistle him down. My mistake! Harry is coming on hot and hard and he's not taking my stop whistle, but he will take his flanks—Go left! Go right!—like a car with failed brakes, all hope's in the steering. The sheep swerve but do manage to get through the fetch gates. I try to calm myself and when he nears I say, in a vaguely distressed way, "Haaarrry!" and he slows but Harry's been telling me one of his ewes is flaky and now, when he pauses at my suggestion, she bolts. With no help from me, Harry barely heads her but the turn around the handler's post is sloppy. The sheep jig left and right before settling on line for the drive gates. Harry's lost confidence and looks back at me, once, twice, three times, for further instructions but the sheep do go through the gates. The crossdrive runs along the far side of the water obstacle and I can see Harry is wishing he could stop this pad, pad, padding along behind these stupid sheep and just fetch them straight to me. Harry hates my peremptory commands that keep him from doing what comes most naturally. This is an appallingly long cross drive—nearly four hundred yards through four sets of panels—and Harry is fed up with it and beginning to sulk, but I tell him this is urgent, real urgent, and though at the last minute they shy, Harry's sheep hit these panels too.

We've done the drive which Harry hates; next, the shedding which we both hate. The sheep veer wide coming into the shedding ring but seem calm on arrival. I make a gap and call Harry through, turning him onto the proper pair with a "tcch, tcch." Okay. While Harry regathers the four sheep outside the shedding ring, I run to the pen.

I try to force the sheep in and they balk and that same flaky ewe bolts but before I can command him, Harry turns her. Words, commands, any words seem fraught with risk so I simply cluck Harry on until he gets the sheep in the pen. I lay him down with a gesture, bring the sheep out of the pen, back to the shedding ring, the sheep are swirling, swirling. THEY'RE LINED UP and that quick Harry comes through and I don't care if the judge likes our shed or not because it is what we can do here, right now, and my knees are shaking.

We were a point and a half out of the top twenty. At the handler's meeting I skipped, the judge said he'd deduct heavily for not regathering the sheep in the ring before the pen, and he did as he'd promised.

The dog has more at risk out there on that huge field than the man, and the dog knows it. Harry has done his part. Ears laid flat on his head, scraggly white ruff catching the sunlight, that squareheaded, sulky, homely, randy son of a bitch is beautiful.

8

Refinement

In 1876, Portugal freed its slaves, in the Balkans Christians and Moslems were killing each other, Porforio Diaz's revolutionary government took power in Mexico, and Alexander Graham Bell's new telephone thrilled crowds at the Centennial Exhibition. Scandals racked the Grant administration, and on a remote Montana hillside, General Custer and his command were killed by Crazy Horse's Sioux warriors. In that year only one in five American farmers farmed the same farm he'd worked ten years earlier, and thirty chartered banks failed. In 1876, when the Westminster Kennel Club held its first dog show, Americans were entering the third year of a long and painful economic depression.

American promises had never glistened more brightly, nor had rich and poor ever been farther apart. Eleven years after the Confederate surrender, the victors were confirmed in their moral virtue.

America was not what it had been before that war, and if its optimism was extraordinary, its fears were abysmal. The fables it told itself clustered at the antipodes, grandiose or cautionary; Horatio Alger or *Maggie, A Girl of the Streets.*

In New York, dowdy, conservative Knickerbockers guarded society's gates against the newer, rougher money clamoring for admission. The new money did not assert venerable family, nor ancient connections; they were distinguished by extravagance and style.

Refinement—that eighteenth-century English notion—had crossed the Atlantic (first cabin) to seduce a rough and ready democracy. Their bastard combined the affectations of the former with streams of cold hard cash. Manhattan in the 1870's must have been a thrilling, nervous, rather uncomfortable place to live. Its smart set devalued nascent American traditions in favor of rigid new "standards" which served to both identify and/or create the essential lady, the true gentleman, the purebred dog.

In 1876, Arnold Burges wrote the following in *The American Kennel and Sporting Field*:

> Although it is claimed that the dogs of our ancestors were superior to our own, this is undoubtedly an error . . . it is certain that we now have dogs half human in intelligence and bearing in their veins the blood of canine kings; and, as during the last quarter of a century the race has been unquestionably improved, it is still reasonable to suppose that the acme of perfection has not yet been reached, and that qualities are still latent which will be brought out under a judicious scheme of breeding from selected animals.

Gordon Stables displays greater optimism in his 1878 account:

Have dog shows then fulfilled their original premise? They have to a very large extent. Within the last ten or twelve years the breed of dogs in this country has wonderfully improved . . . and I hope and trust that the time is not very far distant when either ladies or gentlemen will be ashamed to walk in the street in the somewhat vulgar company of mongrels.

In *The Animal Estate*, Harriet Ritvo's seminal study of Victorian England's attitudes toward animals, the author concludes that dog shows provided Victorian dog fanciers with self affirmation:

The juxtaposition of arbitrarily established criteria (the major purpose of which was to make judgment possible) with swiftly changing fashions, not only in favorite breeds but in preferred types within these breeds, symbolized a society where status could reflect individual accomplishments and was, as a result, lacking in foundation and in constant need of reaffirmation. As most dog fanciers were, in this sense, self created, so their exploitation of the physical malleability of their animals was extremely self referential. Its goal was to celebrate their desire to manipulate, rather than to produce animals that could be measured by such intrinsic standards as utility, beauty or vigor.

Early dog shows were the product of new power, economic and technological, and were haunted by the fear of losing power—a terrifyingly real possibility in those turbulent times. One day a man might control great railroads, and next year be a bankrupt. Morgan survived and Augustus Belmont, Sr., fretted his way through, but Jay Gould—who'd financed the Union war effort—went under, and countless others fell with him. Prosperous Manhattanites needn't travel farther than Elizabeth Street's teeming tenements to sniff the stench of powerlessness. Any man with wealth and half a wit gave vigorous allegiance to impermeable social distinctions: the more arbitrary the better. The rich man's fancies kept him safe.

♠♠♠

August Belmont Sr. was a financier and political mover and shaker. According to an 1877 account, Belmont

> taught New Yorkers how to eat, how to drink, how to drive four-in-hands, how to furnish their houses, how to live generally according to the rules of the possibly somewhat effete, but unquestionably refined society of the Old World. It is no exaggeration to say that on the whole of this continent there is not another house of which the appointments are as perfect as those of Mr. Belmont's. He is not a mere gastronome, a collector of works of art, or a blind adept of fashion. He is an artist in his household. From the livery of his coachman to the menu of his daily breakfast and the disposition of the knick-knacks on the mantelpiece of a spare room of his country-house, everything is to him an object of sincere artistic solicitude.

Belmont's namesake, August Jr., played on a smaller stage. Like his father, August Jr. was a dog fancier. He became president of the American Kennel Club, which he ruled for sixteen years. He funded the American Kennel Club Gazette from his own fortune and consolidated AKC power by insisting that AKC rules must govern every dog show and dog club in America. Those clubs who persisted in going their own way were cast out. When New Jersey Kennel Club delegate Charles Peshal criticized the AKC, he was sued for libel. Even though Peshal won his case in court, the AKC peremptorily expelled the New Jersey Kennel Club from its membership.

In 1888, William Wade wrote the following in *Forest and Stream*:

> The simple truth is, that those who own, breed and exhibit dogs are the ones to control doggy matters, and do it, they finally will. No amount of cuttle-fish clouding of the real facts in Mr. Belmont's latest style will stave off the inevitable. That doggy interests are to be dominated by any such ragged body as the AKC is impossible Two or three men conclude to get up a dog show club, they get the consent of a few more to allow their names to be used as members of the club and the result is a full-fledged member of the AKC The boss bugler makes himself the delegate. . . . The AKC is, and has always been, a laughing stock in doggy circles. It has always been despised, and has never done anything to remove the bad opinion dog men have of it, and there will certainly be some other governing body here.

Alas, Mr. Wade was wrong.

♠♠♠

At the short Labs' preliminary hearing, New York's Sewart and Kissel acted as the AKC's law firm, and though our observer wasn't awfully impressed, she did note that there were a very many lawyers, all expensively dressed. The USBCC Legal Defense Fund had raised twenty thousand dollars to fight an organization who—like the tobacco companies at the time—did not back down in the face of lawsuits.

We were willing, even eager, to find common ground with the AKC, and we would negotiate everything except the Border Collie itself. By then there were two Border Collie splinter groups vying for AKC anointment as official AKC breed club. One club hoped to protect the dog's working ability within the AKC, the other mostly wanted to show in conformation. Both had proposed breed "standards" to the AKC. We didn't pay them any mind; they weren't making the decisions.

When AKC President Daniels said that their Board sought additional information, we took them at their word. We found a Border Collie expert within an hour of each AKC Director and sent that Director a personal invitation to see our dogs doing what they had been bred to do. Our expert would phone them to arrange a mutually convenient time.

Our experts reached answering machines, and no calls were returned. As always, no American Kennel Club Director was interested in seeing a real Border Collie work.

♠♠♠

Harry was a country dog. He drank water from puddles, and his favorite spring burbled out of an old woodchuck hole. Harry was extremely interested in dead animal parts and during hunting season often brought unidentifiable, disgusting objects home. Harry didn't often wear a collar and was rarely on a leash. I wondered what Harry would think of New York City.

We'd been on the road for a couple days. Harry liked tents, but only tolerated motels. He didn't like noisy air conditioners, and motel water stank of chemicals and it never got dark outside the windows.

We came into New York City in the morning. Big trucks in full roar and cars darting past on left and right. The air smelled awful, and Harry did not think he would live long if the car broke down and he had to walk. He laid flat and didn't look up at all.

But as soon as Harry got out of the car in the heart of Greenwich Village, he dropped his snout to the pavement like a bloodhound. Harry was reading the Sunday *New York Times*, dog edition, turning pages as fast as he could. Within five minutes, he'd scented several hundred new dogs and, like a human at a great big cocktail party, Harry was overwhelmed by strangers. So many dogs to meet, so little time.

I brought Harry into my agent's office. Harry smiled and made himself agreeable, and we left him tied to a desk while we went to lunch.

After lunch, Harry and I visited the Washington Square dog park, a shady sandy area where a dozen dogs romped and played. Manhattan is a rich city, and the dog park had good drinking water (Harry approved) and pooper scoopers and benches for the owners to sit while their dogs played.

A memorial mourned the loss of a dog park regular who had died with her Cairn terrier on TWA flight 800.

The dogs were in good condition, purebreds and pound mutts alike, and the owners did nothing to direct their play. Most were mannerly, except a young Australian shepherd who approached new dogs in the most playful manner and after a brief romp tried to mount his playmate. Dogs like to be mounted by strangers about as much as humans do.

When this city dog came over to the country dog, the dog conversation went something like this:

City dog: Hiya, hayseed. This is the most exciting park in New York City which is the capital city of the world. Wanna have a good time?

To which Harry replied, clearly, "You must think I was born yesterday. Son, where I come from you're not even a dog."

City dog's tail drooped and he went away and in a few minutes finally stirred up a fight which the park dogs joined in, but Harry virtuously ignored.

Harry stretched, spoke briefly to two or three fine bitches, drank his drink of water and we set out for home.

Harry told me New York City was real interesting, especially the bitches. But the moment we turned onto our dirt road, Harry was sitting up peering through the windshield, eager to be home on the farm. Because he is a dog, Harry doesn't fear clichés. Harry told me that New York City was a fine place to visit, but he wouldn't want to live there.

♠♠♠

Many dog fanciers believe nothing but good can result from breeding dogs for dog shows, that nothing is lost in the process.

When I first arrived in Highland County, twenty-five years ago, I helped Uncle PeeWee Stephenson shear his sheep. Uncle Peewee was an old-timey forest farmer who turned his ewes out into the woods in the spring and gathered them in the fall. His old-fashioned ewes were so small I could pick them up and carry them to the shearer, but damn they were hardy.

Our ewes weigh 180 pounds, wean twin lambs, and require constant attention. Anyone who breeds livestock understands that every breeding decision is a tradeoff. There are no win/win breeding decisions.

Certain that dog show breeding would swiftly disperse the complex of genetic abilities that made the Border Collie such a useful stockdog, we sought scientific proof for our claims. When Jasper Rine invited me to visit the Dog Genome Initiative, I jumped at the chance.

I was a keen but ungifted student, and the three days I spent at Lawrence Berkeley Laboratory gave me headaches. Guided by Rine, and his compatriot Elaine Ostrander, through the scientific theory and apparatus mapping the dog genome, I'd focus as hard as I could for two or three hours until I couldn't take any more in, "I'm sorry. Not another word. I have to take a nap."

I liked these scientists. They were on the cutting edge, and interested in the whole truth. At a party at Jasper's, young post-docs grilled me for hours about Border Collies, what do breeders breed for, what "behaviors" might be reliably passed on.

♠♠♠

In response to ABC's 20/20 show, the AKC sent a strange, rambling apologia to every member of every AKC dog club, blaming reporters for errors of commission and omission. The AKC saw itself as the helpless victim of an unsympathetic media.

The December 12, 1994, *Time Magazine*'s cover story focused on AKC breeding practices: "An obsessive focus on show-ring looks is crippling, sometimes fatally, America's purebred dogs." The article explained that the "focus on beauty above all means that attractive but unhealthy animals have been encouraged to reproduce — a sort of survival of the unfittest. The result is a national canine-health crisis, from which few breeds have escaped." They quoted Michael Fox: "The best use for pedigree papers are for housebreaking your dog." When they claimed the AKC is perceived as overbearing, AKC Vice President John Mandeville replied, "I think it's a fact of life that people have that fear, and it's unfortunate."

Another fact of life is that, on the heels of the article, the American Kennel Club's Board of Directors voted ten to two to recognize the Border Collie. I believe they blamed us for *Time's* story and took revenge.

December 14, 1994 was not the worst day of my life. There had been worse days.

♠♠♠

The Newfoundland water test took place in calm Pennsylvania lake waters. The test—which involved the dog's retrieving its favorite water toys, towing a canoe about thirty feet, and pulling a life-jacketed victim about the same distance to shore—was proving too difficult for most of the Newfoundlands; some were remarkably poor swimmers. One of the Newfie people informed me that "old-fashioned" Newfies were used by the Canadian coastguard and French lifeguards to save lives, sometimes in very rough waters. I asked why they didn't breed to these dogs. "Oh, but then we'd never get our championship," my informant replied.

Many Newfoundlands have defective hearts. Their genetic heart disease can be discovered by a simple and relatively inexpensive test: it costs about $25.

Every year the Newfie club holds its "National Specialty," a dog show for Newfies only where their "best of breed" is selected. Couldn't they, I asked, solve this genetic problem by hiring a veterinary cardiologist to examine the competing dogs? "Refuse to award a championship to any dog with a bad heart," I suggested.

I was rewarded with incredulous stares.

♠♠♠

That the AKC is the de facto government of America's dogs, and that its decisions are reached, without public input, in secret: this information makes the courteous journalist yawn.

But buying a puppy that gets expensively sick and lives a short miserable life: that's a consumer issue and becomes news.

In response to my *New York Times* Op Ed piece, Randi Locke of Long Island wrote, "As the owner of an arthritically afflicted German Shepherd whose temperament restricts her exposure to family and friends, our dog is a shining example of victimization by the breeding industry — gorgeous, symmetrical coloring, regal posture; an ancestral tree worthy of any European monarch. And an irreversible deteriorating genetic disorder that drives my husband and me to tears."

Time opened its article with the story of "Jake," a golden retriever puppy who has osteochondritis, dysplasia, severe allergies, dry skin, a poor coat, and has started seizuring. "He's a medical mess," his owner said, "It just breaks my heart because he wants to play like a puppy but he can't."

Although there are many reasons to deplore the American Kennel Club, the consumer issue is where the AKC becomes newsworthy. So why didn't they pretend a bigger interest in genetics? Why didn't they disqualify dogs who possess known, detectable genetic defects from receiving AKC Championships? If they were to do so, the cloak of public indifference would once again conceal them, and they could continue ruling the dog world unchallenged. They'd swell their coffers, buy more dog art, assure cushy pensions for loyal staffers.

But instead of adopting this obvious ploy, they exposed themselves to antitrust attention, IRS reclassification, and very bad press. They wouldn't make the most trivial concessions. Were they indeed so stupid they couldn't see their own best interests?

♠♠♠

During the latest interregnum, we'd identified congressmen with farming and ranching backgrounds and matched them to their Border Collie constituents. Within a month of AKC recognition, we fired off a mailing particularly targeting the livestock subcommittee of the House Agriculture Committee. Stephen H. Taylor, New Hampshire's Secretary of Agriculture, protested to the AKC, and worked to bring our case to the floor at the State Ag Secretaries annual convention. The National Federation of Sheep Producers condemned the AKC takeover.

But we didn't have confidence in these protests. Congress wasn't going to act unless and until we could stir up the media, which we couldn't do until we sued. We found ourselves exactly where we hadn't wanted to be — facing a legal fight between a twenty-nine-million-dollar organization and a twenty-nine-hundred-dollar dog club.

We believed that those who historically had bred, registered, and protected a dog breed for years should have some rights over what that dog becomes, that they had some legal interest in that dog's reputation. But if there were any law protecting us, our lawyers couldn't find it. To go to court, we needed a strong legal case and a skilled pro bono attorney to fight for us.

It is neither easy nor pleasant to ask professionals to work for free. I used the beggar's endless courtesies.

I spoke to people I hadn't spoken to since I was a boy back in Montana. One important lawyer turned us down, noting, "It's too bad you didn't bring this up a couple months ago. New York's former Attorney General is a close personal friend. I could have asked him to intervene. Too bad he lost the election."

One Border Collie owner's brother headed a prominent New York law firm. "I haven't spoken to Herbert for ten years," she announced grimly. "But, damn it! It's for the dogs!"

Some who aided our search were AKC renegades who saw us as a safe way to get back at their overlords; most were ordinary ranchers and farmers who wanted to keep on working the dogs they always had.

We must have talked to twenty firms — wherever we had a personal contact (the doggy sister, the nephew who was nuts about sheepdogs). Although they weren't willing to commit to the fight, our cause intrigued some lawyers who gave us hours of free advice.

John Pegram, a top trademark lawyer, asked me, "What about the *Mink* case? It was brought under the Lanham Antitrust Act. Look it up."

A Federal Court had found in favor of mink breeders objecting to someone calling his nonmink fabric "Nutra Mink."

We hadn't been able to trademark "Border Collie" — precisely because everyone knew what the dog was — a working sheepdog. But AKC recognition would *change* what the dog was, as it acquired (without payment or permission) the Border Collie's reputation, and, as the lawyers say, "all the good will appended thereto."

Sheepdog trainer Penny Tose's father was the distinguished Federal Appeals Court Judge John Minor Wisdom. Allen Black of the Philadelphia firm Fine, Kaplan and Black had clerked for Judge Wisdom. The firm had antitrust experience, and their Melissa DeLisle might be willing to represent us, pro bono. Fine, Kaplan and Black weren't afraid of the AKC's high-dollar lawyers. "Send me all your information," Ms. DeLisle said.

I did. After reading the *Mink* case, she tracked down a newer, better precedent. When General Mills started marketing their Pringle's brand of canned "potato chips," the trade association of potato chip manufacturers objected because, unlike traditional potato chips, Pringles were made from dried reconstituted potatoes. How could General Mills call Pringles potato chips? The traditional chippers won.

♠♠♠

Ms. Delisle would represent us pro bono. We'd pay witness, copying, and Fed Ex expenses. If, at any time, Fine, Kaplan and Black became convinced our case wasn't likely to prevail, they'd resign.

We'd be expected to produce expert witnesses—a top geneticist and an economist to prove that the AKC's actions would cause financial damage.

And we'd need to bring all those with a claim to the name "Border Collie" into the suit: all three registries.

Arthur Allen was in his nineties, formidably cranky and sole proprietor of the oldest registry—the North American Sheep Dog Society. Guessing Arthur would be hardest to persuade, we planned to approach him last, after the other two registries had signed on.

Our Defense Fund legal committee consisted of lawyers Penny Tose and Eileen Stein, and myself. For months we'd email each other daily. We were of one mind.

Jasper Rine was willing to be our expert geneticist without charge, and Penny asked the retired chair of the University of Montana's Department of Ag Economics to testify about economic consequences.

Our $25,000 Defense Fund began to look like more money. Once we filed suit and revved up the publicity machine, we could raise more.

In a final effort to avoid litigation, I wrote Judith Daniels that every time our phone rang and "AKC" was mentioned, my wife Anne bought herself a new pair of Birkenstock sandals, and that Anne was in danger of becoming the Imelda Marcos of the Birkenstock set—couldn't we settle this without going to court?

The AKC counsel wrote back. He had no sense of humor.

The principal registry, the ABCA, quickly signed on for the fight. I phoned Senette Parker, acting head of the American International Border Collie registry (AIBC), who said her

directors might have some "trouble" assigning the name
Border Collie to the USBCC for purposes of the suit. I wrote
back the following:

21 August 1995

Dear Mrs. Parker,

*When last we spoke you said the AIBC directors might find that
assigning the name "Border Collie" to us (which we would
promptly license back to you) might be a "sticking point." Since I
didn't explain things very well on the phone and since the AIBC
assignment is central to any successful fight to keep the Border
Collie out of the AKC show ring, I thought I'd try to be more clear.*

*There are a number of organizations, both national and regional,
which use the name "Border Collie." Until recently, the name has
referred to a particular sort of stock dog bred exclusively for its
herding abilities. Now the American Kennel Club (and its surro-
gate breed clubs) mean to begin showing the "Border Collie" in the
show ring. We believe such showing will inevitably lead to breeding
for show ring blue ribbons and will create, at best, two distinct
breeds of dogs both bearing the name Border Collie. We further
contend that the AKC action will cause consumer confusion and
financial harm to the breeders and registries who have protected the
dog's working ability and reputation.*

*I cannot tell you how many hours we have spent locating a law
firm that can put that last paragraph into language which will
make sense to a judge who knows or cares nothing about dogs, or,
worse, presumes (as one New York judge recently did) that "The
AKC are the dog experts." We found a firm, Fine, Kaplan, and
Black, who are experts in antitrust law, more than a match for the*

AKC's attorneys, and they will represent us free. Our $25,000 legal fund can go for their expenses.

I believe that they can force the AKC to either (a) change the name of the dog they register or (b) drop the whole thing.

However, to successfully bring legal action, the USBCC (or someone else) has to represent those presently using the Border Collie name correctly. While we needn't bring in every state Border Collie Association, we must be able to speak for all three registries.

The only practical way for us to do that is for the registries to assign us the right to the Border Collie name so we can legally represent the Border Collie in court. If we cannot get the registries to agree to do that, our lawyers WILL NOT take the case any farther and will probably decide they will no longer represent us. Since the lawyers are working for no pay, they won't continue unless they have a reasonable chance of winning and unless we (and they) represent all the major original Border Collie groups they cannot hope to prevail.

In brief: if the Border Collie registries don't join us, our fight against the AKC will be lost.

For nearly four years we have been hunting a legal strategy that could stop the AKC and a legal team capable of making the strategy work. Strategy and lawyers are in place. What we need now is the approval of the registries to act for the Border Collie.

If we don't get to court or if we lose, the assignment is meaningless and we will return it to you. If we go to court and win, we will assign the name free to any group who wishes to breed the Border Collie as it has always been bred. And, of course, the AIBC will already possess its license.

We have no desire to interfere in any way with the registries. Your policies are your policies, your revenue is your revenue, your standards for merit registration are your standards. We are perfectly willing to guarantee this in a form that suits your lawyer.

Herbert Holmes, President of the ABCA, has agreed to sign on for his organization. If you will sign on, I will approach Arthur Allen last.

We have lined up law, lawyers, and expert witnesses. We can send the AKC packing. And more than that, we will have protected our brilliant, brave, irreplaceable Border Collies.

> *Sincerely,*
> *Donald McCaig*
> *Vice President*

In 1980, when the AIBC broke off from Arthur Allen's NASDS, the new registry became the creature of its secretary, Dewey Jontz. Jontz's son in law, Dean Kaster, took over after Dewey died. I didn't think the AIBC Directors had met in years, and when I asked Dean Kaster who they were he had to think pretty hard. Like the NASDS, the AIBC had become a small rural business, turning out begats papers for a fee. But Dewey Jontz and Dean Kaster had strongly opposed AKC recognition, and Senette Parker, widow of the AIBC's first president, had visited AKC headquarters to protest the takeover.

When Dean Kaster died and Ms. Parker picked up the reins, AIBC affairs were a mess, people weren't getting the papers they'd paid for months ago, and Ms. Parker told me she was awfully busy trying to sort things out. Re the lawsuit? She would consult with her directors and get back to me.

♠♠♠

At the Bluegrass trial in June, everyone wanted to know what was happening with the legal case and what they could do to help, but we couldn't tell them anything because Melissa DeLisle had asked us to keep mum until the suit was filed.

It was very hot, and the outrun was very long. Gael had gotten too old to trial, so I'd retired her. Harry, my only Open trial dog, was having difficulties.

A horseman was spotting sheep—keeping them quiet in one place until the dog arrived—and when Harry got to the top end, his sheep hid under the horse's legs, and Harry couldn't peel them off. Although the same horseman had put out sheep for the National Finals where Harry had run so well, perhaps Harry had never really noticed the horse. After the last run that evening, as dark was falling, I walked Harry to the top end so he and the horse could get reacquainted. Although the horseman had worked all day, he took time to help me make introductions.

The second day, Harry managed to get his sheep off the horse, but that was all he did. Harry was sluggish, wouldn't take his commands, and we didn't even finish the drive.

The AKC swallowed the Cavalier King Charles, and we were getting frantic calls from Jack Russell terrier and Anatolian shepherd breeders.

Some years before he ran for office, Congressman Bob Goodlatte came to a book signing where we had chatted at length. Afterward, he ordered each of my books as they were published.

It was the spring after the conservative revolution. Congressman Goodlatte and I were on opposite ends of the political spectrum but that sort of thing bothered me less than it had in the '60's.

I hadn't expected to spend much time with the Congressman but hoped I could connect solidly with one of his staffers. Some weeks before, a Nebraska rancher had been on CBS, talking about his Border Collie, how useful he was, and I included that video tape in our press kit.

The U.S. Congress was in flux. Newt Gingrich was in charge, his conservatives were going to change everything, and surviving liberals were keeping their heads down. Goodlatte's staffer pointed at the ponderous door of the House Appropriations Committee. "Rostenkowski used to run that like a private fief," she said with some satisfaction. She also said she sometimes thought about becoming a writer.

"Keep your day job," I said.

♠♠♠

The caucus room where we met was a throng of congressmen, staffers, and lobbyists. It was too noisy to talk, so Congressman Goodlatte ushered me upstairs to the house visitor's gallery where we found seats down front.

I asked him how he liked being a congressman. I said I didn't think I'd be able to stand it.

He said it was hard on his family — traveling between his district and DC so often he couldn't even buy his child a dog — but it was wonderful to be here.

Where the action is, I suggested.

He said it was exciting to be part of great changes.

I explained our problem with the AKC, explained how it affected livestock producers, said we'd be perfectly willing to accept the results of a binding vote on the matter. I said we didn't want a lawsuit but one seemed inevitable. We were trying to bring this to the attention of the livestock subcommittee of which he was a member.

He said several members had mentioned the Border Collie/AKC conflict, but we weren't really on the radar screen yet. He asked how I felt about the fight.

I said I wasn't cut out for this kind of thing. I'd rather be working my dogs.

We talked about life and work. The Congressman promised to look over our materials and hoped he could help.

Two other congressmen contacted the Defense Fund, but we were unable to make enough noise to bring the issue before the committee.

The AKC caught wind of what we were doing, and Judith Daniels sent the AKC's lobbyist to Capitol Hill to mend fences.

After the AKC's lobbyist visited, Congressman Goodlatte called me. He hadn't known the Border Collie had once been an AKC "miscellaneous breed," nor that some Border Collie owners wished to see it recognized. Why can't anyone register with whichever registry he or she prefers?

I said they had no moral right to take over a useful dog and turn it into a useless one, that the AKC was a monopoly that didn't pay taxes, and that we'd be happy to abide by the result of a binding referendum.

Hmmph, he said. He said he'd try and arrange a meeting between us and the AKC in his office.

Which was the last I heard from Congressman Goodlatte.

♠♠♠

Sheepdog names are usually one syllable, sometimes two. At great distances a dog can hear a shorter name better; he'll hear his name in the teeth of the storm. Names are pedestrian and traditional: Gyp, Roy, Shep, Gael, Mack, Dot, Bill, Joe, Zip, Bill, Pip. One top sheepdog handler names all his dogs "Glen."

Show dog names are long and frequently witty. A Purina ad celebrated the Basset Hound Ch. By-U-Cals Razzle Dazzle; the Dalmatian, Ch. UKC Gr. Ch. Touchstone Dealer's Choice; the Greyhound, Ch. Shazam's The Journey Begins; the Pointer Ch. B'Wines Jonathon Sundown and the industrious Sheltie, U-Atch, U-CDX AM/CAN Ch. Autumn n' Shenandoah's Murphy, AM/CAN CDX & TD, MX, PT, VCX, TT, TDI. The nineteenth-century fancier's aversion to two dogs bearing the same registered name has produced some odd monikers.

In sheepdog culture, one thinks of the dog before the man. One thinks of Wiston Cap, and only afterwards his brilliant trainer and handler, Jock Richardson, and I'd have to go to my books to find who Wiston Cap's breeder was. One thinks of Templeton's Roy, before picturing Templeton himself; Wilson's Roy, before Tommy Wilson; Dodie Green's Roy, before Dodie; Red Oliver's Roy, before the unforgettable Mr. Oliver.

When Westminster publicist Thelma Boalbay spoke at the Dog Writers banquet and didn't recall the names of dogs that had won previous Westminsters, I had thought Ms Boalbay troubled by faulty memory. I was wrong. She forgot individual dogs and remembered individual handlers because the dogs didn't matter.

The Dog Show precedes the Dog.

I suppose most dog fanciers honestly believe that way back in dog show prehistory, breeders lived with, worked, and studied their breeds before adopting "standards" that exemplified the ideal dog of that breed.

It is unfortunate that this neat logical progression wasn't what occurred. What did happen was that people got together — some who knew a great deal about dogs, some practically nothing — and began showing dogs. The experience was so agreeable they institutionalized it and adopted standards to make the dog show judge's work easier.

Show dogs are the more or less interchangeable *raison d'être* for a very human event, a gala, a chance to flaunt one's credentials and snub those without them. Those black-tie parties — they aren't adjuncts to dog shows — they *are* dog shows.

The Westminster Kennel Club bills its dog show as "America's second oldest sporting event," and the American Kennel Club vows to promote "The Sport of Dogs."

There is precedent for calling hunting dogs "sporting dogs." In the nineteenth century "sport" and "hunting" were nearly synonymous, and when a gentleman said "we had great sport" he was talking about hunting.

Extending this terminology to cover all dogs is troublesome. Whatever its virtues, a Chihuahua is hardly a sporting dog. Surely the Westminster Kennel Club's dog show is an exhibition, certainly a competition, but a "sporting event"? Is the Houston Fat Stock show a "sporting event"? Is there a "Sport of Cows"?

But if the Dog Show precedes the dog, the "sport of dogs" becomes intelligible. The gala, dogs, black tie, snobbery, Madison Square Garden, secrecy, artificial and arbitrary discriminations—that's the "sport of dogs."

♣♣♣

And, save for some bad luck, the "sport of dogs" would have been no more than another human vanity.

That bad luck was the upward mobility of the American middle class after World War II. The American Kennel Club never sought to become the de facto government of dogs, it fell into the job. Since they had become the dog epicenter, the public conferred on them an expertise (which, it must be admitted, they were in no hurry to deny). Without knowing or caring about anything but dog shows, the AKC became "the dog experts." Because nobody else had any coherent program for American dogs and the AKC were "experts," they collected the dog taxes (begats checks). It's not difficult to conjure up some sympathy for the AKC: they were only in it for the parties.

If sheepdoggers had had similar bad misfortune, we wouldn't have done much better. Like the AKC, we are indifferent to everything but one preoccupation: we would glorify stockwork; and scorn the beautiful and useless, the lapdog, the untrainable, the barking watchdog, the dumb but lovable family pet.

The historical fluke that gave the dog fancy and its regulatory body, the AKC, dominion over American dogs has had unfortunate consequences, the worst of which is that no better dog government has emerged to help our dogs survive the rough ordeal of American life in the twentieth-first century.

A better dog government would be democratic (one dog owner—one vote) and represent all dogs, not just show dogs. It would honor a free and critical press. It would devote more of its revenues to the well-being of dogs, and less to expensive quarters and bloated executive salaries. It would be dog savvy and would use genetic science to produce healthier family pets.

Many gated communities ban dogs. New residents of "retirement homes" usually must give up their aging pets, often to the animal shelter which kills them. Dog are not allowed in shopping malls (which means they must stay home or hazard heat stroke left in the shopper's car). By law they are excluded from restaurants and food markets.

I am often invited to read from my dog books at high schools and universities. "Harry, who? You mean you want to bring a dog?" Every weekend, a different neighborhood is featured in the *Washington Post*'s real estate section, Residents are invited to say what they like and dislike about their neighborhood, and charts are provided for typical apartment units: amenities (pool, weight room, etc) and whether small pets (cats, birds, fish) are welcome, and/or a dog. The "cats allowed" symbol is a cute cat face, "dogs allowed" is a silhouette. There are rows of cute cat faces. Sometimes there is one dog silhouette, usually none. If we weren't lucky enough to live deep in the country on our own farm, we could not keep our dogs.

There is no effective government of dogs to protect dog owners or dogs.

In Denver, it is illegal to keep bull terriers and bulldogs, and dogs from those breeds can be killed by law officers on sight. I know tax paying American citizens who dare not drive through Denver with their dogs.

Charles Dickens' "Two Dog Shows" displays a brilliant, intuitive connection. Once we divide dogs into two classes, purebred dog show aristocrats and mongrels, we have reduced all dogs to whims ("fancies") or disposable curs.

Inbreeding show dogs is genetically a terrible idea: the method show breeders employ to produce champions inevitably concentrates genetic defects. The mongrel makes a healthier pet.

Every week, my county newspaper runs photos of shelter dogs which will soon be put to death. While I am grateful the paper does this, I wish I could skip those pages. "Please Take Me Home," the ads say.

Because the AKC is the de facto government of dogs, and because the AKC is obsolete, the laws that most affect America's dogs are often made by those who fear them, sentimentalize them, or are only interested in their suffering. Dogs and dog owners endure an ever more dog-hostile culture without a champion.

♠♠♠

Harry and I were not doing well at the trials — he worked reluctantly, even sullenly. Years ago, when I bought Harry's mother, Gael, in Scotland, I asked John Angus MacLeod (who knew her breeding) what problems I could expect as Gael got older.

"She'll become dour," John predicted.

Was her son Harry becoming dour? He'd always been sulky, but previously, I'd been able to jolly him out of his sulks.

Miss Ethel's sheep are flighty, fast Barbadoes, and I hoped the challenge would interest Harry. Trial dogs need experience on different kinds of sheep: sound ones, flighty, stubborn. The trial dog must understand sheep that are worked by dogs daily, as well as the wild range ewe, who last saw a doglike creature when a coyote was eviscerating her lamb.

At Sunnybrook, Harry ran better than he did at home. It was almost like old times. Afterwards, Miss Ethel said she was doing a little demonstration at an SPCA event in Leesburg, would I come and help?

It was a nice day in a small city park. There were all kinds of dogs, proud owners. A team of Malamutes pulled a wheeled training sled. Harry spotted sheep, and Ethel's Tess came and worked them.

Miss Ethel had sent condolences when Jack Ward's wife died. She'd heard that Jack was going blind. Another fellow (Son? Nephew?) was guiding Jack Ward by the elbow.

Miss Ethel said that she wished Ward was still AKC chairman, that we could have worked something out, that James Smith had been shockingly rude to her over the phone. "Who is Jon Apogee! Who is Jon Apogee!"

Blind men's eyes don't see *nothing*, they see inward. "You may not believe this," Jack Ward said, "But we are all doing our best for the dogs."

No, I didn't believe it. But how could I say so to this grand and shattered king?

♠♠♠

In the months after the takeover, James Smith was replaced as Board Chairman but those directors who voted for us also lost their seats.

When Miss Ethel wrote the new AKC Chairman, Dr. Robert Berndt, Berndt replied that show Australian Border Collies he'd judged had come straight from working sheep and been cleaned up for the show ring. Men think what they want to think and disregard the rest.

The AKC Directors rejected both aspirant Border Collie clubs' proposed breed standards in favor of a Border Collie standard written by the Australian Kennel Club. To reduce the AKC headquarter's appalling Manhattan rent, Judith Daniels suggested that headquarters be moved to North Carolina where their computer registrations are located.

Thus, Judith Daniels lost her job. Ms. Daniels fatally underestimated the importance of those black-tie parties.

♠♠♠

I wasn't getting anywhere with Senette Parker and the AIBC. Although she was perfectly courteous, Ms. Parker did not return my calls and wouldn't answer yes or no. We'd learned that Ms. Parker was Chief Financial Officer of a company with a thousand employees. She was clearly an able business woman, and her silence was baffling. We asked Texans who knew her and her late husband to intervene, and she promised them she would join us — after she talked to her directors. To regain momentum, we called the aged and ailing Arthur Allen, who promptly signed on. In February 1996, the first Border Collies would be shown at Westminster. We were getting desperate.

21 November 1995

Dear Mrs. Parker,

I am pleased to tell you that attorneys for both the ABCA and the NASDS are presently working with John Pegram on an agreement which will permit the USBCC to seek a certification mark and, that failing, pursue legal action under the Lanham Act against the AKC.

I was surprised and pleased Arthur Allen decided to go along with us but he has a young sharp attorney who understands how doing nothing will damage all the Border Collie registries.

I am also surprised that AIBC has not yet signed on. I know that you agree with us that AKC registration will harm the Border

Collie. Since most of the obedience people currently registering with AIBC will opt for AKC registration, you must see how the AKC move will damage your own registry.

Yes, I know you are exceptionally busy and have no AIBC deputy but surely negotiations can be handled by your attorney. If you have him contact John Pegram or me, we can begin talks.

I hope you have a pleasant Thanksgiving.

Donald McCaig
USBCC Legal Defense Fund

9

Harry

Harry, Harry-Harry,
Harr-Harry, HarryHarry . . .
(like the Hari-Krishna chant)

We had a chance to save our breed from the ignorant clutches of a multi-million dollar oligarchy. We had a modest war chest, a little guy vs. big guy media story, and a first class pro bono attorney. All three registries needed to sign on if we were to bring suit. The North American and the ABCA were willing, but Ms. Parker didn't return my calls, didn't answer my letters, and wouldn't name the AIBC lawyer so our lawyer could talk to him.

In desperation that winter, I made reservations to fly to Houston to meet with her face to face. She called to say her mother was in the hospital, so I cancelled my trip. In the early summer, I made another reservation and she called to say she had a family member in the hospital. I sympathized but said I was coming anyway and should we meet at the hospital or her home?

The AIBC lawyer finally called. He seemed reasonable and willing to discuss joining the suit. When I called our lawyer, she was about to go on maternity leave, and since the AKC had been registering Border Collies for a year and a half without legal protest, our case was now too weak.

So that was that. We returned the money we'd collected for the Defense Fund and used the residue to fund free eye clinics.

We'd lost. If I had been more persistent earlier. If I'd been ruder and flown to Houston. If I'd been able to persuade Senette Parker to join us . . .

I had betrayed my dogs.

♠♠♠

Heather Nadelman, a scholar of American religious movements, thought the Dog Wars were very like a religious war, and her insight changed my thinking. Suppose AKC directors didn't bother to come see the Border Collies whose fate they were deciding, not from smug ignorance but because they believed they had already "seen" them; to them, Border Collies were a breed, like any other—perfectly capable of being judged in a show ring and occasioning black-tie parties. What difference did it make to the dog show religion that Border Collies also worked sheep? There are no sheep in a show ring. Many AKC directors were also all-breed dog show judges, high priests of their faith. Who were we to instruct them about any dog?

I do not imagine anyone at 51 Madison Avenue ever analyzed the AKC's motives for breed takeovers. It must have seemed their right to acquire more breeds, with or without the acquiescence of those who presently "had" them, because after all, weren't they the dog experts? Weren't they the government of dogs? Those people who criticized them, even fought them — weren't they cranks and crackpots who'd never belonged to the Westminster Kennel Club, had no status whatsoever in the show ring, and if invited to a black-tie dinner (Heaven forbid!) might show up in Levis and dung-smeared cowboy boots.

Sometimes the AKC seemed to believe it was defending fanciers who ardently wished to show Border Collies in the AKC show ring, although there weren't ten of those people in the contiguous 48 states. Perhaps they thought they were "rescuing a breed" from the uninitiated and that the dog might have its proper type identified, and the once-humble Border Collie might become "refined."

In part, they were motivated by bureaucratic imperatives. At our first meeting in 1991, Louis Auslander let slip that someone (Directors, staffers?) had visited the UK Kennel Club, learned the KC registered Border Collies, and wondered, "Why can't we have them too?" Perhaps the AKC was embarrassed the dogs that were winning all their obedience competitions weren't "theirs," but they consider non-show ring dog activities as nonessential — important only as buttresses for the show ring.

When AKC staffers argued with traditionalists that they should abandon their venerable snobberies and recognize every breed they could, the staffers were just doing what staffers have done since the time of the pharaohs: increase their importance by swelling their organization.

Many dog people believed, "They're just in it for the money. They want those extra registrations."

I couldn't agree. If the AKC was uninterested in our dog's working abilities, they seemed no more interested in money; indeed, they reacted as if insulted whenever we mentioned the filthy stuff (they never brought it up). The mailings they sent out, the genetics program they started to counter what had become a public relations nightmare, cost them far more than they'd ever earn from Border Collie registrations.

The AKC took a tremendous risk for inconsequential financial gain and damaged their reputation in the process. The Dog Wars surfaced issues they had no wish to surface, created new critics, and emboldened those who'd previously been intimidated into silence.

The fight cost people their jobs. Many who orchestrated AKC strategy during the Dog Wars no longer work at 51 Madison Avenue.

For a long time I believed the AKC was uniquely institutionally stupid, but Ms. Nadelman's observation that the Dog Wars was like a religious war was a better explanation.

When private language like "I've run in 300 trials," and "Australian Border Collies come to the show ring straight from working sheep" are accepted without demur, when the AKC doesn't insist on inexpensive genetic tests before awarding its coveted championships, religious explanations become persuasive.

Throughout the fight, I kept stumbling over a simple truth without quite seeing it: dog fanciers and their creature, the AKC, really do believe that what is most valuable about any dog can be judged in the show ring, that the show ring is the sole legitimate purpose and reward of all dog breeding. They even believe, against all evidence, that the show ring "improves" breeds.

By the secular standards that inform ordinary judgments, this belief is absurd. But if the dog show is seen as an object of faith, secular standards are irrelevant. Certainly the AKC's faith in the show ring is no more implausible than the fourth-century creed I recite every Sunday in the Williamsville Presbyterian Church.

We see show dogs as useless, they see our dogs as "unrefined." We and the AKC understand each other as well as fundamentalist Christians understand the Dalai Lama.

♣♣♣

After the smoke settled, Border Collie owners and the AKC both lost the Dog Wars. The AKC has never registered as many Border Collies as they hoped to: 2,000 annually, a tenth of the dogs the ABCA registers. While they have twice the "herding" events as sanctioned traditional trials, ordinary citizens seem to understand what's real and what's not, and while big sheepdog trials draw ten thousand spectators, AKC herding events are insular and invisible.

The Dog Wars and our public criticism of the AKC's dog savvy and motives may have taken some luster off "AKC Reg." and "AKC Ch." pups. After all, unless you're interested in dog shows, AKC pups are no better pets, they're likely to be more expensive, and may be less sound than mutts from the animal shelter. The moral satisfaction of adopters who rescue their pet from almost certain shelter death, and the diminished stature of purebred dogs have cut AKC registrations by 40 percent since the Dog Wars, and I am happy to say that the AKC's virtual Border Collie is widely and popularly known as the "Barbie Collie."

Though it is not as powerful or credible as it was, the American Kennel Club remains a threat to the Border Collie. They suck up sponsorship money and have more cash and political clout than they know what to do with. They're smarter than they were during the Dog Wars and more media savvy.

Our breed's strongest defense is the farmer and rancher's need for useful — not AKC-titled — sheep and cattle dogs. Without sheep, the breeding, training, and keeping of sheep-dogs loses its rationale. Other once useful dogs like bulldogs, deerhounds, and wolfhounds are either extinct like the Welsh Gray sheepdog and the Dalesman, or they became dog show dogs — faux replicas.

Sheep numbers in the US have declined from 53 million in 1942 to seven million today. Much of the best western sheep range has been purchased by billionaires and turned into elk and buffalo preserves.

On the East Coast, sheep shearing is a dying profession, and the wool clip just covers shearing costs. Ordinary farmers are turning to hair sheep.

Some recent statistics are encouraging. There's been a ten percent increase in breeding ewes in the last two years.

As we had expected, when the Border Collie was recognized, confusion proliferated. Jon Katz, who has to my knowledge never trained any sheepdog to a useful standard, has become a popular Border Collie "expert." Ten years ago, almost all those who hung out their trainer shingle were Open handlers or visiting Brits. Since the AKC began its herding program, dozens of trainers have popped up. Many have never trained a dog to Open level, never competed against the best, and never learned what a sheepdog can and cannot do.

♠♠♠

One sheepdog handler had a dog, Ben, a Border Collie with Bearded Collie whiskers. A friend invited her to an AKC "herding trial."

"It'll be fun, you'll see."

When she got to the trial grounds, she jumped Ben out to empty and went to inspect the course. Tremendous ruckus: "LOOSE DOG!" "LOOSE DOG!"

Ben's owner had no idea she'd caused the hullabaloo until her friend ran over to demand she leash Ben. "I'm so embarrassed," the friend told her.

Ben didn't have many difficulties with the AKC "A" course—after all, the pen's the exhaust. When his handler came off, she took Ben to water. "LOOSE DOG!! LOOSE DOG!!!"

Her friend pretended they were strangers.

As it happened, Ben won the trial and the judge, a professionally coiffured, pigeon-shaped woman, congratulated Ben's owner and dipped into the barrel for Ben's prize toy.

Ben had never seen a dog toy before so naturally he killed it. Within instants, the toy was scraps of rubber and cloth.

Appalled, the judge drew back, lifted her snoot and pronounced, "Oh dear! We don't know how to respect our toys, do we?"

That silly woman, who doubtless believes that every dog must have dog toys, and that "Borders" or "B. C.'s" (as she probably calls them) get "shown in herding" when they're not good enough to be "shown in breed," that woman is not—repeat not—the Border Collie's greatest enemy. Yes, the dog fancy is dangerous. As Iraq taught us: resolute ignorance and great power is a terrible combination. While we can't do much about the AKC, we sheepdoggers can do better for our dog than we have in the past.

Wannabe prophets should be humble. Remember the "DOW: 40,000" prediction just before the techno-bubble burst? Remember Y2K and those believers who reserved every Jerusalem hotel room December 31, 2000—because Christ was supposed to return one minute past midnight the first day of the second millennium and they wanted to greet Him personally?

But it's one thing to assume that tech stocks will continue to rise indefinitely, or that the Book of Revelation contains, er, *revelations*. It's much easier to identify consequences of long-term trends.

The Border Collie isn't North America's premier stockdog because of our unselfish stewardship, nor our ineffectual attempts to promote the breed. What preserves the Border Collie is its value to ordinary farmers and ranchers. Only when a breed becomes a luxury can silly folk dominate a breed and determine for what and how it is bred.

In the next twenty-five years, the Border Collie should be affected favorably by trends that will challenge most everything else.

American agribusiness famously requires more than one calorie of energy to produce one calorie of corn. Global warming, skyrocketing Chinese and Indian energy demands, declining oil reserves, and wild fisheries will bring severe droughts and the end of cheap energy, water, and protein.

Necessity has always been the Border Collie's friend. Since sheep (and goats) are adapted to low energy rearing on marginal land, what is ruinous for agribiz and confinement rearing should serve sheep and sheepdogs—despite the likely demise of the goosedog industry. (When protein gets expensive enough, people won't chase geese, they'll eat them.)

When plastics cost more, wool might even be valuable again.

These same predictable conditions will affect our trials. As sheep and goat flocks increase in numbers there should be more trials, and "for-profit" trials may coexist with traditional hosted trials. The days of the behemoth RV are numbered. When gas hits $10 a gallon, we'll be pulling dog trailers behind eentsy teensy little cars. We won't be able to travel as far or campaign as hard as we do now. Regional finals will replace today's national trailer race and who knows, maybe regional teams will share a bus to the Nationals.

♠♠♠

Does the AKC want our registrations? Sure. It annoys them that such an important breed isn't truly "theirs." If they had all the Border Collie registrations, they'd have the tenth most popular dog in America, and doubtless they'd triple the ABCA's $200,000 annual income. That ain't peanuts.

But dog fanciers have no status among farmers and ranchers. The Wyoming rancher who bought a stockdog because its sire went "Best of Breed" at Westminster would be laughed out of the Stockman's Saloon.

The AKC will continue to poach ABCA registrations (unless new identity theft laws can prevent them). But considerable ABCA income (and many breeders' incomes) derive from dogs that compete in AKC sanctioned agility, rally, and obedience. In the face of those profits, there's no political will to stop the poaching.

No Border Collie organization wishes to compete with the AKC in agility or obedience, and we certainly won't offer conformation titles. I expect we'll keep on doing what we do best and leave other dog sports to those genuinely interested in them.

There are more agility competitors than stockdog handlers and the popularity of agility, rally, flyball and frisbee will grow. Formal obedience, which is tedious to watch (and untelevisable) will continue to decline in popularity.

But, as the AKC is dominated by a small cadre of dog show people, stockdog culture is dominated by a small number of trial handlers. There's no way agility people will "take over" a registry which is already providing everything they want from it: sound working pups. Take us over? Why should they? They've already got everything they want.

Our dogs have made us look better than we are. As breeders of the world's best and most popular stockdog / agility dog / obedience dog we sheepdoggers have gotten away with being underfunded, ill-organized amateurs.

We have too many organizations and no media and sponsorship expertise. Most of our novice handlers are coming out of the dog fancy. Though we welcome those immigrants, residues of dog fancy culture — the rule bound mentality, the kowtowing to authority (and concomitant resentment), the ugly language and, yes, the dog toys — will inevitably accompany them.

Sheepdog culture is consensual, not legalistic, and beginners succeed by studying the masters, not hiring a top handler to "show" their dogs. It is a genuinely friendly, courteous, extremely dog savvy culture; a much pleasanter culture than the dog fancy. In the next twenty-five years, some of these immigrants will become top handlers and influential in our breed. By then I hope our culture has changed them more than they will change us.

So long as sheep and goat farming thrive there will be a working Border Collie.

We've always needed our dog more than our dog needed us.

♠♠♠

Harry's troubles at trials came from a bad heart valve. When he was stressed, Harry hadn't enough oxygen in his blood to think or attend or even walk straight.

Along with Harry's heart medicine the Virginia Tech veterinarians gave me a stern warning: no strenuous activity, keep this dog quiet!

But Harry was a sheepdog, and sheepdogs are genetic workaholics. Telling Harry "forget it," every time he asked, "We going to work today, Boss? Is this a work morning?" was intolerable.

For man and dog alike, there are things worse than death.

So I used Harry for light farmwork, here and there, bringing in the big flock for feeding, putting them out after they ate, nothing tricky or high stress.

We'd been lambing for two weeks and caught sleep in shifts, and Anne woke me at 2:00 am because a ram had climbed one fence and was destroying another to break through to the ewes. "I heard this twanging!" Anne said.

Harry was our only dog that'd work by flashlight.

Sure enough, the minute Harry and I stepped outside it sounded like a rock group smashing guitars. That ram was hitting the high tensile fence: "KA-WHANG, KA-WHANG."

And before Harry and I could turn him, Ram-O broke through and he and his new girlfriend fled into the night. There were 80 ewes in that field, many due to lamb momentarily, the rest were aged ewes no longer able to bear and rear lambs. Seeing Ram-O disappear was like watching a serial rapist dropping over the wall of a nunnery.

It was pitch black, raining a little, and at the far end of my flashlight beam, I couldn't see anything but sheeting mist. Since dogs (and sheep) see better than humans at night, I switched the light off. "Harry, I need you," I said. When Harry was well away I turned the light back on and headed for the barn lot.

At night you can hear their hooves before their eyes come into sight like bobbing amber road reflectors. Sheep are frightened by the unknown and by flashlight the familiar gateway looked dangerous to them. Harry kept working back and forth, turning on a dime just to hold them in the opening. When I switched off my light, the gateway looked sheep-friendlier and they bleated through.

Gate cutting sheep is what it sounds like, I let a few slip through and slam the gate on others while Harry kept up the pressure behind. I had an eye peeled for Ram-O (he's the big sheep with horns). The braver ewes go in first, next the medium brave until only a dozen timid ewes and Ram-O were still in the lot and Ram-O charged the gate which I hold (barely) by throwing all my strength to it. Harry came to my rescue and peeled Ram-O away.

I latched that gate and ran to our cutting chute. The sheep'd been through these chutes a hundred times, but by flashlight they looked safe as a slaughterhouse ramp, and Ram-O was getting cross. Ram-O weighed three hundred fifty pounds, had horns like a bighorn mountain sheep, and didn't care to have any other animal stand between him and the objects of his desire. Horned rams can smash a dog against a fence, break bones, even kill him, and horned rams are not susceptible to reason. Ram-O lowered his head, pawed the ground. Harry slid forward, crouched like a predator. Ram-O backed a step, another, bumped into a ewe. Oops. He charged. Harry jumped out of the way and nipped Ram-O's hock and got right back in his face. Ram-O backed up and lowered his head. Harry showed many many teeth. One inch at a time, lit by my flashlight, Harry backed that huge animal into the chute.

When I closed the gate, it was over: ewes in one pen, Ram-O back with his pals. I was half whupped, and Harry was panting hard. We'd been at it for an hour.

Back inside the house, Harry lapped a little water. The vets say the first sign of Harry's heart getting worse will be coughing. When his lungs fill with fluid, he'll try to cough it out.

I told Harry to get up on the bed next to me and laid my hand on his silky flank. A few minutes later he was asleep. It took me somewhat longer to fall off, listening to my dog's breathing.

Appendix A

Correspondence

June 7, 1991

Louis Auslander
Chairman, Board of Directors
American Kennel Club

Dear Mr. Auslander,

I am writing you because of what we have in common. Both of us have been lucky enough to have owned at least one great dog. Pip, my first Border Collie, taught me how to work stock, showed me what he saw, taught me his limits and graces and showed me some of mine too. He inspired me to write *Nop's Trials*, which has made me friends all over the world. I can never repay the debt I owe to Pip.

Although we had a working sheep farm, when I bought Pip I knew nothing about herding or Border Collies and had someone asked me back then would I like an AKC herding dog or non-AKC, like most naive Americans I would have said "AKC" and I would have been wrong. Not because you guys are bad guys or don't care about dogs or didn't have a great dog who got you started, same as I did. But most Americans want pet dogs and that's what the AKC registers: purebred quality dogs that get along pretty well in most homes. Sure, dog owners who wish to work their dogs have AKC programs including, recently, a herding program, but the show ring is dominant in the AKC and you can't judge a show dog without a conformation standard and dogs bred to a conformation standard lose their herding ability within a few short generations.

Perhaps you don't agree with me. Perhaps you envision a "Dual Purpose" herding dog.

For a hundred and twenty years, important Border Collie sires and dams have all been working dogs, chosen by their success in standardized sheep dog trials. There never has been a conformation standard for Border Collies and unregistered dogs that do well in trials, in Britain or the U.S., may be registered on merit. My most recent breeding was to Tommy Wilson's Roy — a dog that came over from Scotland, unregistered, and was registered in the U.S. on merit.

This single-mindedness has produced a dog that is a brilliant herder but not a terribly good pet. One of the effects of British Kennel Club recognition was a Holocaust for Border Collies when pet buyers discovered this interesting new breed and the dogs ended in animal shelters by the thousands.

As a working dog, however, none of the "Dual Purpose" breeds can compare with it. In the hundreds of sheep dog trials I've seen — trials open to any dog, any breed, registered or unregistered — I've seen one imported Beardie, one Kelpie, and I read that a Belgian Tervuren won a novice trial somewhere. All the other dogs and all the good ones were Border Collies.

When the AKC started its herding program, you came to the Border Collie people to help—not because we're the nicest people in the world (Bob McKowen rightly calls us smug)—but because we're the only people in the country who know very much about herding dogs. June 15th and 16th, Ethel Conrad, our President, and I will attend the AKC Herding Clinic at Frying Pan Park. We'd like to help you work out the glitches in your herding program.

You may honestly believe that imposing a conformation standard on the Border Collie will not diminish its working ability. I assure you, in the dogs I've seen in Australia and Great Britain, it certainly has. Mr. A. Philip Hendry, Secretary of the ISDS (the parent British registry) recently wrote me that show and working Border Collies are now distinct breeds "as different as chalk and cheese."

If our dogs are lovely, dog politics seem unlovely. I wish you would tell me why the AKC is pursuing the Border Collie with such unseemly avidity. Surely the AKC isn't after our puny registration fees. Surely an organization we have rightly admired for so many years wouldn't stoop so low.

I would estimate that fully 95% of Border Collie owners oppose a conformation standard and full AKC recognition. Nor, I hasten to add, are we interested in either the SKC (States Kennel Club) or UKC (United Kennel Club). If you asked me for the name of a single reputable, well known Border Collie person who wishes to join the AKC, I honestly couldn't help you. The group your staff created out in Kentucky is about as convincing as the puppet government Saddam Hussein installed in Kuwait.

Please tell why the AKC wants to register a breed its responsible owners fear will be damaged by such recognition. Please tell how AKC recognition will help Pip.

I am obliged to tell you we Border Collie people are unified and will fight any attempt to hijack our dogs. Our lawyers say we've got a good case and we'll fight politically too. I often write for National Public Radio, *The Washington Post,* and *The New York Times*, and I can assure you this fight will make the AKC look bad. You seem to be encouraging frivolous, irresponsible dog breeding. The AKC appears greedy, antidemocratic, a bully.

I don't want that, nor, I'm sure, do you. What we Border Collie people want is the status quo: the Border Collie in the AKC Miscellaneous Class, able to compete in obedience but not herding.

Some years ago, your then President, Bill Stiffel, came to a trial here in Virginia and on his return, he said, "The Border Collie should never be recognized, or shown in the breed ring." I hope you will agree with him, for the sake of all the great dogs—yours and mine.

Sincerely,
Donald McCaig
Vice President, USBCC

July 10, 1991

Mr. Donald McCaig
Yucatec Farm
Williamsville, VA 24487

Dear Mr. McCaig:

This is in response to your June 7, 1991 letter. Thank you for bringing this matter and your concerns to my attention. The point you make that we have something in common because we must each have owned at least one great dog is well made. It also struck me as ironic in that the dog of mine which immediately came to mind is a Whippet. I say ironic because at its meeting on June 11, 1991, AKC's Board of Directors approved regulations for the newest AKC perform-ance event, which is lure coursing, something my particular dog and every Whippet I've ever known seems to be particu-larly enthusiastic about.

If you will permit me, I would like to give you my reaction to one major theme in your letter, which concerns the Ameri-can Kennel Club and performance events. In a nutshell you appear to have adopted the "little lie" (or to be a little softer small distortion) point of view, "little lies" which are re-peated and repeated, and just because they have been fre-quently repeated in some quarters begin to be believed. In this case the "little lie" has to do with AKC and performance events, or more broadly AKC's attitude toward performance dogs.

AKC strongly and unequivocally supports the idea of performance dogs. We believe dog breeds were created to perform a function. We believe in supporting activities that emphasize a breed's working qualities. Sure, it's true that AKC has been the foremost proponent of conformation shows. We have also been the foremost proponent of working retriever trials, Beagle trials, obedience for all breeds, and numerous other performance events.

My point is not to enumerate the performance events AKC supports but to emphatically tell you AKC believes very fundamentally that part of its mission is to encourage and foster the working ability of dogs.

I am uncertain exactly what the impetus for your letter has been. Obviously recent developments with the Australian Shepherd must have something to do with your concern. Let me briefly summarize the Australian Shepherd situation. The Australian Shepherd has had *no* recognition by the American Kennel Club. This means the breed, unlike Border Collies, is not in the Miscellaneous Class, and therefore cannot participate in any AKC events. Suffice it to say the American Kennel Club was approached by representatives from the Australian Shepherd fancy. They wanted AKC recognition for their breed so they can participate in our events. I would add that the representations from the Australian Shepherd fancy were made by a much larger number of people than almost all other breeds we have ever admitted to the Miscellaneous Class. AKC's Board has decided to recognize the Australian Shepherd.

There is another aspect of our recognizing the Australian Shepherd which has to do with the Miscellaneous Class. The Miscellaneous Class is intended to be a transitory class for breeds moving from non-recognition by the AKC to full recognition. The Miscellaneous Class was never intended as a "safe harbor" for a breed, only to take advantage of AKC obedience trials. We are taking a close look at the Miscellaneous Class, the breeds in it, and how it is used or misused as the case may be. I can tell you AKC's Board of Directors is evaluating the Miscellaneous Class. The Board might well decide it does not want breeds that want only to use the class as a parking place in order to take advantage of obedience competition. I can also tell you it is just as likely the Board will decide to leave the status quo in the Miscellaneous Class.

I want to emphasize I enjoyed your letter to a point. Obviously you are concerned about your breed and its working abilities. I admire and respect your position. Frankly, however, I find your statements about your lawyers saying you have a good case and your threats involving the media distasteful and uncalled for. I state unequivocally that the AKC would never have any problem communicating its purpose, mission, and deeds. AKC stands for the protection of purebred dogs and we have a record of accomplishments unparalleled anywhere in the world.

The United Stated Border Collie Club, at least as represented by you and your letter, believes strongly in your breed and your purpose. I can assure you the American Kennel Club believes every bit as strongly in its mission and purpose. We are always willing to discuss matters of serious interest to serious fanciers. I offer you and the representatives of the United States Border Collie Club an open invitation to sit down with members of our staff, and myself if you would like, for a full, open candid discussion.

Sincerely,
Louis Auslander
Chairman, Board of Directors
American Kennel Club

♠♠♠

July 25, 1991

Ms. Beth Miller
10748 Patterson
Durand, Illinois 61024

Dear Ms. Miller:

Thank you for sending your thoughts about not wishing to have the Border Collie recognized as a breed by The American Kennel Club, making it possible for Border Collies to be registered with AKC.

I do not know of any movement on the part of AKC to recognize the Border Collie. I have no idea what the future may hold. Additionally, I have little to say about it.

I realize your letter was written with the best of intentions and that what you wrote is what you believe. Without going into the Border Collie issue at all, I do have to tell you that I resent people slamming The American Kennel Club as "ruining" breeds. That is very far from the truth. My own background in dogs is an excellent repudiation of that accusation which seems to be the mainstay of your leadership for a number of years. I think you should know that the largest Border Collie club was comprised of obedience people from AKC events.

I owned the leading sire in the history of the breed in German Shorthaired Pointers, and he was the equal in the field of any Border Collie in his line of work. In addition to having 54 field trials wins, including wins against American Field Pointer and Setters, he also won 12 specialty shows including the national. And I can refer to thousands of dogs that were aided in the field by AKC participation rather than being ruined as you and some of your fellow Border Collie people seem to believe or make accusations to that effect.

I wish the issue involving working dogs was as simple as you indicate in your fifth paragraph. It is a little more complex than that, as it is also with the Border Collie.

I also ask how you know the dog you bought is a Border Collie. You have a piece of paper but does it look like a Border Collie? Of course it looks like a Border Collie because it has an appearance standard.

I also want to point out that there is nothing keeping the Border Collie out of the pet market. It receives probably more publicity than any other breed in this country. It is publicity that drives people to the market. Look at the great number of (Chinese) Shar-Peis that have been sold in this country without being registered with The American Kennel Club.

I know you didn't mean it this way but I do take some offense when I read a statement that says, "The majority of breeders wish to maintain our breed as a working dog without interference from the AKC." I refer to the word "interference." I'm not aware that AKC has done anything but allow the Border Collie to participate in obedience trials with an ILP number for almost 28 years. I don't think that's interfering.

If you were aware of some of the things that your leadership is doing at the present time you might have second thoughts about your own organization.

As I stated previously, AKC is not soliciting the Border Collie as a breed. However, AKC is giving greater consideration to all of the non-AKC breeds that have been requesting acceptance.

I hope you will accept my letter in the same spirit which I accept yours.

Sincerely,
Robert H. McKowen
Vice President, Performance Events
American Kennel Club

October 29, 1991

Mr. Donald McCaig
Yucatec Farm
Williamsville, VA 24481

Dear Mr. McCaig:

This will acknowledge receipt of your letter of October 7, 1991. I wish to thank you and your associates for coming to our offices to discuss the future of the Border Collier and its continuation in the American Kennel Club's Miscellaneous Class.

There are two things I do wish to restate so that there is no misunderstanding of our position. If the Board of Directors of the American Kennel Club in the future decides to change the status of the Miscellaneous Class in any manner and should said change affect the Border Collie, we will certainly advise you *after* the decision has been made and *before* it is implemented. In fact, we will do so for all clubs in the Miscellaneous Class.

Your second paragraph contains the following sentence — "If either financing or fairness is part of your discussion, surely we can negotiate something satisfactory to both of us." There will be no special financial consideration affecting our decision. You made a similar statement at our meeting and you have again alluded to it in your letter. I trust the aforesaid clears up any false impressions you may have had as a result of our meeting.

May I say in closing that I speak on behalf of the Board of Directors and convey the policies as they exist at this time. I cannot speak for my successors or future Boards who have the responsibility and the duty to make policy as they see fit.

Sincerely,
Louis Auslander
Chairman, Board of Directors
American Kennel Club

♠♠♠

AKC: Hands Off the Border Collie!

We own Border Collies. Our dogs are companion dogs, obedience dogs and livestock herding dogs. For hundreds of years, Border Collies have been bred to a strict performance standard and today they're the soundest, most trainable dogs in the world.

The AKC wants to push them out of the Miscellaneous Class and into the show ring. They seek a conformation standard (appearance standard) for the breed.

We, and the officers of every single legitimate national, regional, and state Border Collie association reject conformation breeding. Too often, the show ring fattens the puppy mills and creates unsound dogs.

We will not permit the AKC to ruin our dogs.

Arthur Allen, President, North American Sheepdog Society

Sharon Anderson, Trainer, OTCH

Chris Bach, Numerous Gaines Placements

Bonnie Barry, 2 OTCH, Gaines & World Series Placements

Robert Barlow, President, American Border Collie Association

Bill Berhow, Winner, National Sheepdog Trial Finals, 1989, 1990

George Bernard, Vice President, American Boarding Kennel Association & CKO

Winnie Bigelow, OTCH, Numerous Honors

Ethel Conrad, President, United States Border Collie Club

Gail Dapogny, Two World Series Wins

Janice DeMello, 3 OTCH, Gaines Wins

Eric Engberg, News Correspondent

Bruce Fogt, Winner, National Sheepdog Trial Finals, 1987

Jerusha Gurvin, 2 OTCH, Numerous Honors

Margie Gibbs, Trainer

Sally Glei, OTCH, Gaines Placements

Bob Griner, #1 Obedience Border Collie in U. S.

Kay and Dick Guetzloff, Kennel Ration Dog of the Year, Gaines Winner

Barbara Handler, OTCH, Obedience Judge

Vicki Hearne, Trainer. Author of *Adam's Task*

Debbie Hotze, 2 OTCH, Gaines and World Series Placements

Pat Kaiser, 1st Border Collie OTCH in US, Gaines Winner

Dean Kaster, Secretary, American International Border Collie Registry

Jack Knox, Trainer, Teacher

Charles Krauthammer, Syndicated Columnist

Janet Larson, Founder, Border Collie Club of America

Sandra Ladwig, 2 OTCH, Gaines and World Series Placements

Janet Lewis, OTCH, Numerous Honors

Donald McCaig, Author of *Nop's Trials*

Nathan Mooney, President, U.S. Border Collie Handler's Association

Helen Phillips, OTCH, Obedience Judge

Lewis Pulfer, Winner, National Sheepdog Trial Finals, 1985

Ralph Pulfer, Winner, National Sheepdog Trial Finals, 1988

Mike Randall, Captain, Champion U.S. Flyball Team

David Rogers, Winner, National Sheepdog Trial Finals, 1986

Clint Rowe, Trainer: *Down and Out in Beverly Hills, White Fang*

Strobe Talbott, Editor-at-Large, *Time Magazine*

Shirley Tipsward, Agility

Hazel Thompson, 2 OTCH

J. C. Thompson, Agility

Mary Whorton, 3 OTCH

Liz Wilson, OTCH, Numerous Honors

Joanna Yundt, Obedience Judge

June 16, 1994

Ethel B. Conrad, President
United States Border Collie Club
Sunny Brook Farm
Rt. 1 Box 23
White Post, VA 22663

Dear Mrs. Conrad,

I am sure you are aware that a significant number of Border Collie owners have expressed interest in taking advantage of The American Kennel Club's numerous competitive events and services. Since 1955, the Border Collie has been in the Miscellaneous Class, limiting the events that titles may be earned in, to obedience and tracking. There has been no movement of the breed toward AKC registration since entering the class. Full recognition would allow your dogs to compete in herding trials, agility, and conformation, just to name a few of the areas that would become available to this versatile breed.

My purpose in contacting the United States Border Collie Club is to determine if you have an interest in achieving full recognition and participation in all American Kennel Club activities.

Please respond to this office by July 29, 1994. If no response is received by that date, I will consider that the United States Border Collie Club has no interest in pursuing full registration, or becoming the AKC parent club for the breed.

Please feel free to call me if you have any questions. I look forward to hearing from you soon about this important matter.

Yours truly,
Linda C. Krukar
Project Administrator, Dog Events
The American Kennel Club

♠♠♠

STAND BY YOUR DOG

In its June meeting, the American Kennel Club's Directors voted 11-1 to change the Border collie into a show dog. They are seeking a group willing to act as breed club for this new dog, produce an AKC-approved conformation standard so AKC judges can judge it, apply for full AKC recognition, and turn over registration revenues for AKC staff salaries, medical insurance, perks and pensions.

That the great majority of Border Collie people all over the country despise this idea doesn't bother the AKC. They are big and powerful and accustomed to getting their own way. There are at least two tiny clubs clamoring to give the AKC what it wants.

Since 1975, the United States Border Collie Club has sought good relations with the AKC. We cherish the Border Collie's right to compete in AKC obedience and tracking and hope to see the dog able to compete in AKC agility as well.

We are willing to work for fair AKC compensation for the dog's participation in these events. As this is written, we are seeking a meeting with the AKC Chairman and staff. We'd like to come to a meeting of the minds and avoid a knock-down, drag out fight.

The AKC is not invincible. At present, the AKC is troubled by an eleven million dollar lawsuit, media interest in allegedly fraudulent AKC registrations, and questions from a federal agency. We are willing to add to their troubles. We are prepared to mount a major media campaign, move against the AKC politically and sue both the AKC and its putative breed club. The AKC has no legal or moral right to change our dog.

We are joined in this fight by all the major Border Collie organizations, the top Border Collie obedience and herding people. Very few of those who keep Border collies as companion dogs wish to see them controlled by the AKC. If you join us, we can win.

5 WAYS TO STAND BY YOUR DOG

1. Write, phone or visit an AKC Director. Let him (her) know how you feel about your dog.

2. If you have local media contacts, let us know. We'll help you get our story on the air.

3. If you know a congressman, tell him how important the Border Collie is to the American livestock farmer. Ask why

the AKC—an organization that no longer protects purebred dogs—should continue to be tax exempt?

4. If you are willing to protest at a major AKC event, let us know. Perhaps the Westminster Kennel Club Show could use some excitement.

5. And, of course, send cash. We expect that every dollar of our Defense Fund will go to the legal battle to come. If you can't decide how much to give, ask yourself how much your dog is worth to you, then divide the figure by ten. Geneticists say ten years is how long it'll take them to change the dog forever.

June 29, 1994

To whom it may concern:

I am writing to advise against the notion of trying to breed for a standard conformation in the Border Collie because of the virtual certainty this action would have on severely reducing the working qualities of the breed. My arguments are based on principles of genetics, molecular biology, and DNA recombination that have been solidly established for more than 50 years. My qualifications for offering this advice are rooted in 18 years of genetics research, 12 years of university level teaching, service on governmental advisory committees and having published approximately 70 research papers in various aspects of genetics in leading research journals. In addition, I served for three years as the Director of the Human Genome Center at Lawrence Berkeley Labs.

As background information, it is important to remember that complex traits in any organism, including dogs, are typically controlled by the combined action of multiple genes scattered throughout the genome. In other words, there is no one gene controlling conformation just as there is no one gene controlling the ability of dogs to work stock. Another important piece of information is that there are two copies of every gene in dogs, one copy from the maternal lineage and one copy from the paternal lineage. The products of the two genes are often identical, in which case an animal is said to be homozygous at that locus. However, in many cases the genes differ in subtle but important ways. In these cases the animals are said to be heterozygous, and the two forms of the gene are referred to as alleles. In cases in laboratory organisms like fruit flies in which the role of many genes are known in considerable depth, it is possible to effectively breed for remarkable combinations of traits. However, I want to make it absolutely clear that there are likely to be on the order of several hundred to a thousand genes involved in controlling behavior and morphology of dogs, and we have yet to learn the identity of a single one. Therefore, any rational discussion of the genetic manipulation of these traits has to be grounded firmly in solid genetic principles and guided by whatever precedent is available from related studies.

If one were to breed dogs for a fixed conformation, in all probability one would be breeding for homozygosity of many different genes that contribute to conformation. By this action, all other genes that are near the genes controlling conformation would become homozygous as well due to genetic linkage. Only one dog in hundreds would be recombined within a million base pairs of any of the genes affecting conformation. Thus hundreds of nearby genes would become homozygous. Unfortunately, which alleles of the nearby genes that would become homozygous would be determined by chance since, from first principles, the alleles of any two loci can be either coupled or in repulsion. If the alleles of a gene controlling a desirable behavior were linked to, but in repulsion to, alleles of a gene that determines a "desirable" conformation, then by breeding for the conformation allele one is destined to breed away from the allele for the desirable behavior, with that behavioral allele at risk for extinction. If the coupling relationships of the two genes were in equilibrium, the loss of the desirable behavioral alleles would not be absolutely certain. But in practice, linkage disequilibrium is more common and would be especially more so in animals that have evolved as recently as dogs. Thus based upon these considerations alone, the working Border Collie as currently known could easily be lost to efforts to breed a uniform conformation.

In fact, there are several other reasons that breeding for conformation could breed for loss of working qualities, but I shall discuss only one more here. There are well established examples from human genetics and from genetics of other creatures in which there is an advantage to being heterozygous at a locus over being homozygous for either allele. There is no reason to believe that the same would not be true for at least some genes controlling behavior. By breeding for a conformation standard that "breeds true," by necessity any heterozygote advantage for linked loci would be lost, risking the loss of allelic combinations that may have been selected for by Border Collie breeders. Put another way, it may be nearly impossible to breed for a particular behavior based on heterozygous advantage and still achieve a homogenous conformation.

In summary, any change in Border Collie breeding that would lead to the development of a conformation standard would place the unique behavioral capabilities of this breed in severe genetic jeopardy. It would be difficult for a credible scientist to declare something as being absolutely impossible. However, armed with our present meager knowledge of the genetic basis for behavior and morphology, the notion that one could achieve a standard conformation for Border Collies and maintain their working qualities is simply foolish. The experience in other breeds in which the field trial dogs and the show dogs become genetically non-overlapping groups provides a common sense example of how, in genetics, selection for one trait usually comes at the expense of another. In principle, it may some day be possible to learn enough about the genes that would be desired in the show arena and the genes that would be desired in the field and combine them into the same dog. In fact, my own current research may help the process along.

However, even the most optimistic estimates place that time decades away. I believe it unlikely that any reader of this letter will live that long.

Sincerely,
Jasper Rine
Professor of Genetics
Department of Molecular and Cell Biology
University of California, Berkeley

Appendix B

The Collie Standard

Published in 1871, Benjamin Jowett's *Dialogues of Plato* was the first Greek-to-English translation in sixty years and emphasized Plato's logic rather than his mysticism. This new Plato seemed familiar to common-sensical Victorians. What do we mean when we use the word "table" if not a real object which resembles more or less well the ideal "table"? Aren't our real-world tables imperfect examples ("Platonic shadows") of the ideal?

And living, breathing dogs—are they not slightly imperfect versions of the ideal foxhound or greyhound, setter or collie?

John Henry Walsh was a London physician, keen sportsman, and editor of *The Field*, the most influential hunting and kennel journal of his day. An early bench show enthusiast, Walsh believed that

> In all breeds of dogs which are useful to man there are certain attributes which are essential to the full development of their powers in the right direction *and by these attributes it is easy to estimate any animal of the breed under consideration* [emphasis mine]. Thus a greyhound must have a form calculated to

develop high speed and for distances averaging somewhat less than a mile. A foxhound should have speed also, but united with high powers of scent and stamina sufficient to carry him at a speed somewhat less than that of the greyhound for ten times the above distance.

Despite some niggling — how can one judge speed in a fifty-foot show ring? — Walsh's logic is appealing. A Chihuahua can't catch a hare because the dog's legs are too short. He can't pull a sled because he hasn't enough bulk and he'd freeze to death in arctic weather. Similarly, a mastiff is unlikely to be much use as an earth dog because he can't fit himself into a fox's burrow. In such extreme cases, morphological dog standards make sense: a no-nose pug can't track fugitives as well as a bloodhound can.

But when trivial morphological differences are used to predict performance in morphologically similar breeds, the theory fails. There is no way to use a show ring to predict which greyhound might win races or which sheepdog will be a good working stock dog.

The Dogs of the British Islands, a collection of sixteen Walsh essays from *The Field*, was published in 1867. Later editions appeared in 1872, 1878, and 1886. Illustrated with a "never exhibited" Scotch Colley, Walsh's 1867 account describes sheepdog behaviors in some detail:

No dog has so large and valuable an amount of property entrusted to his care as this faithful creature. Naturally clever and intelligent, he is susceptible — in good hands — of very high training, and his performance is frequently surprising.

. . . Their homing faculty is extraordinary, and it has been asserted the Scottish drovers would send them back from Smithfield to the Highlands alone with a wave of the hand. Whether Scotch or English, the value and skill of the dog depends chiefly upon the temper and intelligence of the shepherd. Some surly, morose, ignorant, discontented men look upon the poor creature as a thing at which they can, in their worst humours, hurl their crook, or which they can use as a safety-valve for their sour humours. These fellows return from every sheep fair with a fresh dog in a string which they soon tell you is as bad or worse than the last; for the dog is a fine judge of character, and, having made out he is in bad hands, he takes up a stubborn, defiant air if he is bold, or becomes in a few hours nervous, shy, and cowed if he is timid. With such a shepherd's dog the flock are either harassed and driven about uselessly, losing condition, and "never looking well" or the dog is of no assistance whatever, and the ill-conditioned owner has all the work to do himself.

The good-tempered sensible master, on the other hand, knows how to control the high-couraged or to bring out the nervous dog. The training of his "fellow-servant" is a pleasure to him, and he gradually gets his dog to the highest possible pitch of training. From an eminence he will gradually ask his dog to attend the wave forward or backward of his hand, and to copy the old dog, or broken dog, which he never sells until the young one has been made "handy." He will teach him the simplest things first — such as to bring in one or two stragglers and then leave them alone; to "lay down" and "keep off" whilst he sets the fold; to keep up the flock like another shepherd, as he walks before the flock along the road to a change of pasture, with his dog behind; then he will divide his flock, and placing the dog in the middle, and going himself behind, he will show the animal how to keep and drive together the flocks of different owners, if required, without mixing them. He will teach him to bark ("speak to 'em," he calls it) at a signal, and by degrees he will get him to sweep round a large flock, perhaps of thousands, a mile away, and, having collected them to bring them as steadily and patiently, and with all the importance of the lawful owner, to his master, and save him so many weary steps, pretending to bite the stragglers but never really using his teeth. By degrees he will become perfect in his work, and in the lambing season will show extraordinary gentleness to the lambs. Indeed, it is asserted that some dogs have been seen to push the weak ones in the direction of the fold, and to steady them when they tottered with their heads. If he is a dog of marked intelligence, he may even be trusted to lie all day upon an eminence and to watch the movements of thousands of sheep grazing below him, for he will keep all in their proper district; and when he hears his master's shrill whistle, he will "go round" and drive them home.

A third- or fourth-rate shepherd will be content with a dog doing very little for him, and he has no idea how to teach him to do more. A really clever shepherd will get his dog to "do anything but carry a hurdle," and will begin the dog's education as soon as he can "head" — that is, go faster than a sheep. He will also put him in the way of doing by artifice what a "blunder-headed" shepherd would never think of. In a narrow lane, for instance, he will get his dog to jump the fence, run down it unseen, and head the flock without flurrying them at all; or, if they get "blocked," and the front sheep will not move, he will teach his colley to run over the sheep's backs, and thus move them on.

At the end of his 1867 account Walsh instructs the dog show judge that "we should give the points as follows: head, 20; temper, 20; shoulders, 10; coat, 10; colour, 10; back, 10; loin, 10; feet and legs, 10." The only difference between the 1867 and 1872 edition is the elimination of "temper" — the dog's behaviors — from the judge's consideration.

Ten years later Walsh (and his readers) were beginning to value dogs very differently. Show champions illustrate the 1878 and 1886 editions, and the text would be familiar to the modern dog fancier. Instead of describing the dog's work on moor and hill, Walsh theorizes at length about the ideal Collie's coat, and he ignores working behaviors in favor of anecdote:

> A curious case which a short time ago happened to myself would almost lead to the belief that the colley understands the meaning of a conversation between members of the human family. Entering the drawing room of a lady who has a celebrated dog of this variety as a pet, I was met by the question, "What do you think of my dog — is he not a perfect beauty?" After looking him over as he lay on the rug, and with a desire to tease my hostess to whom I owed a Roland or two, for her previous many Olivers administered in badinage, I replied very quietly, "Yes, certainly, if he had but a colley coat and a little more ruff." The words were hardly out of my mouth when the dog rose from his recumbent position,

seized one of my feet in his mouth, gave it a gentle but vicious little shake, not sufficient to scratch the leather of my boot, and then lay down again. There was no emphasis on my part, and not a word uttered by the lady until after the act was completed, when I need scarcely say that eyes and tongue told me that I was rightly served. Anyhow, it was a remarkable coincidence; but from a long knowledge of the dog I really am inclined to believe that G— knew I was "picking holes in his coat" and resented the injustice accordingly.

Not only has the Collie been detached from the work he was bred to do, he is now supposed to be as concerned about his coat as dog fanciers are! It is a short distance from Walsh's anecdote to the modern fancier's commonplace "Of course he still has herding instincts. He herds our children and the cat."

In the 1878 edition, the standard was elaborated. Walsh assigns 10 points to the Collie head "which resembles that of the fox, should be wide between the ears, tapering towards the eyes, which are in consequence set rather close together. The top of the head is flat, and there is little or no occipital protuberance, and a very slightly raised brow; but the facial line is not absolutely straight. The volume of brain is considerable, and the skull looks smaller than it really is, in consequence of the amount of frill in which the occipit is embedded." The first American edition of Walsh's influential book was published in 1879. American fanciers probably never saw Walsh's earlier description of Collie working behaviors.

The standard in the 1886 edition was written by the Colley Club, who thought the head (and expression) was worth 15 points. "The *skull* of the colley should be quite flat and rather broad, with fine tapering muzzle of fair length and mouth the least bit overshot, the *eyes* widely apart, almond shaped and obliquely set in the head; the skin of the head tightly drawn, with no folds at the corners of the mouth; the *ears* as small as possible, semi-erect when surprised or listening, at other times thrown back and buried in the ruff."

In 1893, when Czar Nicholas sent fifteen of his Borzois to Victoria, they were crossed with the show collie to produce the narrow-snouted, genetically unsound animal we know today. The modern beast bears only a passing resemblance to the original nineteenth-century "Colley" taxidermied at the Walter Rothschild Natural History Museum at Tring in Hertfordshire. That unrefined dog looks like the Wicklow Collie or a yellow Border Collie.

Walsh wrote the following in the 1878 and 1886 editions:

> In Scotland and in the north, as well as in Wales, a great variety of breeds is used for tending sheep depending greatly on the locality in which they are employed, and on the kind of sheep adopted in it. The Welsh sheep is so wild that he requires a faster dog than even the Highlander of Scotland, while in the lowlands of the latter country a heavier, tamer and slower sheep is generally introduced. Hence it follows that a different dog is required to adapt itself to these varying circumstances, and it is no wonder that the *strains* [emphasis mine] are as numerous as they are. In Wales there is certainly, as far as I know, no special breed of sheepdog, and the same may be said of the north of England, where, however, the colley (often improperly called Scotch) more or less pure, is

employed by nearly half the shepherds of that district, the remainder resembling the type known by that name in many respects, but not all. For instance, some show a total absence of "ruff" or "frill," others have an open coat of a pied black and white color, with a setter shaped body; while others resemble the drover's dog in all respects. But without doubt, the modern "true and accepted colley" has been in existence for at least thirty years.

As a good Platonist, J. H. Walsh believed these regional collie breeds were inferior copies of one ideal. One word: "Colley." Hence one breed, a "noble" dog resembling those owned by Queen Victoria. Walsh's confused notion of different "strains" (modern dog fanciers call them "types") are — alas — still with us today. By conflating distinct breeds dog fanciers describe "show" and "hunting" retrievers, "show" and "racing" greyhounds. Of course, it is the "show" type that comes closest to the Platonic ideal.

Today's working Border Collie is an amalgam of many of the original Colleys Walsh described. Like those hard working dogs, and unlike Walsh's "true and accepted colley," it is a breed because of what it can do, not what it looks like.

The Border Collie standard is performance.